Steal This Idea

Steal This Idea

INTELLECTUAL PROPERTY RIGHTS AND THE CORPORATE CONFISCATION OF CREATIVITY

MICHAEL PERELMAN

palgrave

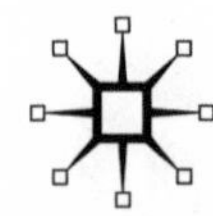

STEAL THIS IDEA

First published 2002 by PALGRAVE™
175 Fifth Avenue, New York, N.Y.10010 and
Houndmills, Basingstoke, Hampshire RG21 6XS.
Companies and representatives throughout the world.

PALGRAVE is the new global publishing imprint of St. Martin's Press LLC Scholarly and Reference Division and Palgrave Publishers Ltd. (formerly Macmillan Press Ltd.).

ISBN 0-312-29408-5 hardback

Library of Congress Cataloging-in-Publication Data

Perelman, Michael.
Steal this idea : intellectual property rights and the corporate confiscation of creativity / by Michael Perelman.
p.cm.
Includes bibliographical references and index.
ISBN 0-312-29408-5 (cloth)
1. Intellectual property—United States. 2. Corporations—United States. I. Title.

KF2979 .P47 2002
346.7304'8—dc21
2001048393

A catalogue record for this book is available from the British Library.

Design by Planettheo.com

First edition: April 2002
10 9 8 7 6 5 4 3 2 1

Printed in the United States of America.

CONTENTS

Hoffman, Abbie. 1971. *Steal This Book* (New York, NY: Pirate Editions).

Back in the 1960s, after thirty publishers turned this project down, Abbie Hoffman's *Steal This Book* became a best seller. Today, the direction of permissiveness to which the title appealed has changed its course. The atmosphere in which Hoffman's tongue-in-cheek suggestion of petty larceny against business would find broad acceptance has given way to an environment in which the law sanctifies giant corporations' grand theft of the fruits of creativity.

INTRODUCTION

HOW INTELLECTUAL PROPERTY RIGHTS ENRICH THE FEW WHILE UNDERMINING LIBERTY, SCIENCE, AND SOCIETY

The New Economy of Intellectual Property Rights

This book explains how the ever-tightening grip of intellectual property rights threatens to undermine science, economic progress, personal liberty, and democracy. Of course, intellectual property rights are nothing new. They are just a fancy name for patents, copyrights, and trademarks—privileges that have been with us for centuries. But in recent years, intellectual property rights have taken on unprecedented powers. Before I go any further, consider the following examples of the abuses associated with intellectual property.

Suppose that you suffer from a rare, but fatal, disease. A corporation offers you a treatment that costs a half-million dollars for the first year and a somewhat lesser amount each year thereafter for the rest of your life. Then you find out that government scientists had discovered the medicine and that a quirk in the law allowed the company to have the exclusive license to market this medicine. This maneuver allowed the company to catapult itself into the ranks of the leading biotech companies.

Suppose that you plant some seeds, then at the end of the harvest season you save some seeds to replant next year. Should you do so, you may well find yourself charged with theft.

Suppose that you are an undergraduate work-study student partially supporting yourself by doing low-level chores in a university laboratory. You come up with some ideas that are unrelated to your duties in the laboratory. You think that these ideas might have some promise, so you patent them. Not too long thereafter, in retribution for your action, you find yourself working on a chain gang having been convicted of misappropriation of property.

I did not have to make up these scenarios. A government report found that Genzyme did actually charge that much for alglucerase in the treatment of Gaucher disease. Monsanto sends out armies of inspectors to check on whether farmers are replanting its seeds. Petr Taborsky did spend time on a chain gang for the nefarious crime of patenting his own idea. Welcome to the brave new world of intellectual property.

I will discuss these examples in more detail later in the book, but they are only the tip of the iceberg. Today, the scope of intellectual property is exploding. Like a lethal virus that quickly mutates into even more deadly forms, claims of intellectual property rights are vaulting into undreamt-of areas. Worse yet, the legal system is validating these claims.

You can patent a human gene, an animal, software code, a way of doing business, and even a number. These and many other forms of intellectual property rights are fast supplanting more conventional forms of property as the principal basis of wealth and power.

The Perversion of Intellectual Property

The original intent of intellectual property rights did have some merit. Supposedly, society granted the monopoly rights associated with intellectual property—and they are monopoly rights—to encourage creative activity. In addition, the law required that the applicant immediately disclose information regarding the proposed patent.

In a comparatively backward country such as the early United States, advanced science and technology comprised a relatively small part of the overall production process. Instead, traditional knowledge remained the basis for most production processes.

The information disclosed in the patent process very likely could have spurred others to innovative activity, especially because inventors were solitary figures. In addition, scientific activity was fairly sparse. Moreover, the patents at the time were narrowly applicable to a particular product or process rather than general discoveries with far-reaching consequences. This characteristic meant that the patent system was relatively unambiguous. Consequently, the sort of litigation associated with the contemporary system of intellectual property would not pose inordinate costs.

Today, matters are completely different. Intellectual property covers just about everything. The system is riddled with overlapping claims. The contemporary system of intellectual property, rather than spreading information, creates a pervasive atmosphere of secrecy. Litigation is becoming far more important than creativity. In fact, I will show that intellectual property rights threaten to stifle creativity. Taking a historical view, we can compare the system of intellectual property to a stimulant that may well have promoted economic and cultural progress in an earlier period, but now threatens to exhaust creative activity.

Even in the arts, intellectual property rights offer very little to the mass of creative artists. In contrast, intellectual property rights grant enormous powers to corporations that distribute music or run movie studios. These corporations typically wield their power to the disadvantage of the artists, as well as society at large.

In science, intellectual property rights encourage secrecy and wasteful duplication of effort. They hold back economic progress by fostering inefficient monopolies. They encourage costly litigation that dissipates an unimaginable amount of time and resources. Over and above these problems, intellectual property rights pervert the entire scientific process by undermining the traditional incentives to engage in the basic scientific research essential to developing future improvements in technology.

I want to put to rest the popular notion that intellectual property rights represent the major vehicle of progress in modern market economies. The same sorts of problems that infect the scientific process repeat themselves in the economy as a whole. So, intellectual property rights, rather than advancing economic development, will actually obstruct future progress by disrupting the scientific process by promoting fragmentation, unnecessary duplication, and secrecy. At the same time, intellectual property rights will bog business down in a morass of horrendous litigation.

Finally, on a larger scale, intellectual property rights shape the very nature of society. Intellectual property rights already constitute an increasingly large share of property in the advanced market economies, especially the United States. Activities surrounding intellectual property are fast becoming the core economic functions in advanced capitalist economies.

Besides the damage that intellectual property rights impose on the scientific process, intellectual property rights concentrate wealth in the hands of the few. The unconscionably unequal distribution of wealth and income in the United States today does incalculable harm, largely crippling the potential of millions of people brought up in a destructive environment of poverty and deprivation.

I will also show how the increasingly powerful hold of intellectual property threatens society in a number of more fundamental ways, including the subversion of democracy itself.

Setting the Stage

Indeed, a major theme in this book will be that, for the most part, the undeniable economic progress of recent decades had little to do with intellectual property. Instead, recent spectacular scientific and technological advancements largely represent the fruits of earlier public investment in science and technology, even though private corporations later won the intellectual property rights. In short, in the field of intellectual property rights, as is the case so often in advanced market economies, costs are socialized, while benefits are privatized.

These changes are political. As one legal theorist explained, "Property is never for long anything more or, really anything different from what some politically appointed court says it is."[1] Although the public at large has had little input in the transformation of intellectual property rights, they affect virtually every aspect of life today, from the food you eat to the kind of medical care that you get. Intellectual property rights largely determine the sort of entertainment that the market offers. More important, expanding the boundary of intellectual property restricts the range of what people are allowed to do. In addition, measures to enforce intellectual property rights have to develop even faster than intellectual property does. As a result, intellectual property rights have a major impact on the personal rights, as well as the civil liberties that people enjoy as citizens.

The promotion of intellectual property rights may well rank as one of the very highest priorities of the current government of the United States. Again, I will explain these claims and give more concrete examples later in the book.

For now, I only want to set the stage for a deeper discussion about the extent to which intellectual property is changing the ground under our feet. What is happening is not accidental. Powerful forces are presently using their enormous influence to change the way society works. I will explain more about this process as you work your way through this book.

When I began my research for this book, I was already under the impression that the government of the United States was engaged in a dramatic construction of a new economic regime based on stronger intellectual property rights. I knew that this transformation was occurring on a world scale through agencies such as the World Trade Organization.

At the time, I had never dreamt how far-reaching this transformation would be. Nor had I understood that debates over intellectual property rights were a recurrent theme in the conflicts about the political economy of the United States.

I realize that at first glance, this orchestrated transformation of the economy might seem reasonable. Many holders of intellectual property rights would have people believe that their good fortune represents an appropriate reward for the sort of individual achievements that power the modern economy. The popular press, insofar as it takes note of intellectual property rights at all, treats them as if they were a natural part of the so-called New Economy, which seems to offer seductive promises of convenience and prosperity. This prosperity, as I noted before, is far from shared by all. Instead, intellectual property rights have contributed to one of the most massive redistributions of wealth that has ever occurred.

Intellectual Property and the Distribution of Income

I will call attention to the connection between intellectual property rights and the widening disparities in income distribution, both globally and domestically. The bulk of all intellectual property is held by a handful of firms, universities, and well-off individuals, accentuating the grotesque inequalities of wealth and income that put a large portion of future generations at risk.

Worldwide, the rich have become richer to an unimaginable extent in recent years. The members of the "Forbes 400," a compilation of the 400 richest people in the United States, have a combined net worth of $1 trillion—greater than the gross domestic product of China.[2] Between 1995 and 1998, the average annual income for a member of this elite group rose from $50 million to a staggering $110 million, while the tax rate on their adjusted gross income fell from 30 percent to 22 percent.[3] The obscene wealth of a Bill Gates of Microsoft or a Phil Knight of Nike, or even of the lesser instant Internet billionaires, alongside the sizable residue of poverty that blights the contemporary United States, reminds us of the existence of a link between distribution of income and intellectual property.

Emblematic of the extent of this new distribution of property, outside of those who have inherited their wealth, three of the four richest people in the world, according to a *Forbes* magazine survey, owed their wealth to Microsoft, one of the major holders of intellectual property rights,[4] befitting the so-called New Economy in which DOS Capital has supplanted *Das Kapital.*[5]

Of the 60 new names on the *Forbes* list, 19 had fortunes derived from Web businesses.[6] While *Forbes* can celebrate the wealth of the few, the poverty of the many means that the world loses the intellectual potential of the vast majority of people, some of whom are undoubtedly among the most brilliant people in the world.

Internationally, a regime of intellectual property rights condemns the poorest countries of the world to an even more disadvantaged future. For example, the United Nations reports that in 1993, just 10 countries accounted for 84 percent of global research and development expenditures. These same countries controlled 95 percent of patents registered in the United States during the past two decades. The rich, industrialized countries now hold 97 percent of all patents worldwide. Compounding the inequity, more than 80 percent of patents granted in developing countries belong to residents of industrial countries.[7] No doubt this situation has worsened in the intervening years.

Protecting the intellectual property of the already rich nations channels the world's wealth away from the impoverished countries. This increasing gap between the rich and poor nations provides the wealthy countries with an even greater advantage in producing intellectual property in the future. This relationship between intellectual property and the polarization of the distribution of wealth and income has yet to register, either in the public consciousness or in the literature of economics. Earlier students of monopoly had no difficulty in making a direct connection between the elevated markups that sellers charged and a corresponding loss on the part of their customers. They realized that the pervasive influence of monopoly harmed the public at large, over and above the obvious losses that individuals suffered as consumers. To my knowledge, nobody has continued to apply the logic of that tradition to present-day moguls, who owe their power to the special monopoly protection given to their intellectual property.

The question of intellectual property rights goes well beyond considerations of abstract measures of the distribution of wealth, such as Gini ratios. Right now, people are literally dying because of the powers of intellectual property rights. For example, in Africa, where AIDS is rampant, transnational corporations have been using the U.S. government to pressure poor African nations to refrain from using their legislative authority to obtain licenses for the right to produce pharmaceuticals at a more affordable price. The drug companies have been charging people many times more for treatment than the majority earn in a year.

Advocates of this new distribution of wealth tend to explain it in terms of the personal achievements of the beneficiaries of intellectual property rights. On closer examination, the development of their intellectual property owed more to public investment than to their personal accomplishments in most of the cases, except for artistic creations. Under this new regime, ideas first developed in public universities or government laboratories, even after they have become common knowledge, suddenly transform themselves into the private property of multinational corporations.

For example, in September 1997, a Texas-based company named RiceTec Inc. won U.S. Patent No. 5,663,484 on basmati rice lines and grain. Of course, this company, a subsidiary of RiceTec AG of Liechtenstein, the chairman of which is reigning Prince Hans-Adam II of that tiny principality, did not invent basmati rice. Farmers in the far-off Greater Punjab region, now divided between India and Pakistan, have been growing basmati rice for centuries. This effort to win an intellectual property right ignored the longstanding contributions of the Punjabi farmers in developing basmati.

Only under intense pressure from the Indian government and public interest groups, and in response to adverse action by the U.S. Patent and Trademark Office, did the company withdraw its patent claims for basmati rice. The Patent and Trademark Office did grant the company three of its claims for varieties of rice that it developed by cross breeding with traditional basmati rice.

I do not pretend that intellectual property is the only cause of the increasingly unequal distribution of income around the world. During the same decades that intellectual property rights have been strengthened, in the United States the rights of unions have been weakened; cheap imported goods have displaced millions of decently paying working-class jobs; and the tax codes have radically tilted in favor of the wealthy. While discussion of these and many other explanations for the growing inequality in the distribution of income have been fairly common, the relationship between the increasing power of intellectual property and the distribution of income has almost gone without comment.

Treating Knowledge as Private Property

I already mentioned that intellectual property rights have a long tradition, dating back centuries. However, this new regime of intellectual property rights is unlike anything the world has ever known. I will show that most of the privileges associated with this new regime are artifacts of a particularly misguided approach to the treatment of knowledge and information.

Ironically, while defenders of intellectual property rights sanctimoniously cloak themselves with the rhetoric of the free-market, in truth, intellectual property rights represent an annihilation of market forces. After all, intellectual property rights are monopolies, a fact that the term "intellectual property" conveniently obscures.

The natural outcome of the present system of intellectual property rights is a world of excessive litigation, intrusive violations of privacy, interference with scientific research, and a lopsided distribution of income. Although defenders of this new regime claim that it will provide incentives that will promote rapid technical and scientific progress, I will show why it will ultimately lead to a morass of litigation and confusion that will inevitably derail the scientific process.

Although holders of intellectual property are quick to demand their rights under the patent system, the initial logic of the patent law was supposed to be an implicit exchange. The inventor would profit from monopoly rights for a short time in return for revealing to the public information about the invention. This arrangement was intended to encourage people to invent and innovate, while spreading information at the same time.

The lengths of time that holders of patents or copyrights get to exercise their monopoly rights today wildly exceed the time necessary to encourage intellectual work. Worse yet, perhaps the majority of these increasingly powerful rights reflect no contribution whatsoever on the part of the holder. Patent holders today often reveal no important information to the rest of society. Instead, people or corporations are holding rights to intellectual property merely on the basis that they were the first to stake a claim, even though they made little or no contribution to the overall discovery.

The impulse to make such claims is not necessarily new. Karl Marx claimed that the invention of the windmill caused the emperor, the nobility, and the priests to squabble over who owned the wind.[8] What is new is the degree to which the legal system has sanctioned such demands. Later, I will go into detail regarding some particularly egregious cases.

In a sense, this aspect of intellectual property rights is not at all novel. After all, throughout history, a small minority of the population has become wealthy by claiming property rights to land or other goods that belonged to the community at large. So you should not be surprised that today, when knowledge and information are so crucial to the economy, the tradition of looting of the commons should extend to knowledge and information.

The government, as well as the mass media, mostly remains silent about the flagrant abuses associated with the current system of intellectual property. Even worse, when either the government or the mass media address the issue of intellectual property, they actively support the alleged rights of those who claim ownership of intellectual property. In this environment, those who protest the unjust claims of those who hold the private property rights to this public property find themselves either ignored or marginalized. This emerging system of intellectual property represents an ominous extension of "the social costs of private enterprise."

I should also mention another dimension of the public support of intellectual property. While a company such as Nike may not have been as dependent on government research as a pharmaceutical producer or an

Internet company, virtually every company that depends on intellectual property owes an enormous debt to the government for protecting its intellectual property, not only domestically, but also internationally. For example, to expand the market for American films, the U.S. government uses its political muscle to get other countries to reduce their support of their domestic cinema. Similarly, the U.S. government has been bullying poor nations to prevent them from making lower-cost versions of lifesaving pharmaceutical products. I will show how this tactic is contributing to the AIDS crisis, especially in Africa.

Rather than engaging in frank discussion about the public contribution to private intellectual property rights, we hear instead endless praise about the efficiency of the emerging system. If this new regime were to become so efficient that everyone could have more because of the great contributions of intellectual property, then the maldistribution of wealth might be an acceptable price to pay. Under that scenario, the benefits of the resulting technological efficiencies could be so widespread that virtually everyone could be better off. Unfortunately, these enclosures of the mind will actually have baleful consequences for technology, as well as the distribution of wealth and income.

I will devote a considerable part of the book to showing how intellectual property rights are undermining the development of future technology by corroding the academic environment, as well as the proper functioning of science itself. Intellectual property rights undermine the working of the market, making the economy become far less efficient. Finally, intellectual property rights create a more intrusive society that corrodes individual rights, especially privacy. I covered the last problem of intellectual property rights in an earlier book, *Class Warfare in the Information Age.*[9]

Presently, the lure of profit combined with the lack of adequate public funding has seduced the scientific community largely to turn its back on basic research and to concentrate on corporate-sponsored research. In addition, within this increasingly profit-oriented research environment, employers are compelling researchers to adopt a code of secrecy, even though open communication has been one of the essential features of the scientific achievements of the past.

While this strategy might produce some quick returns, its long-term implications are ominous. Those interests that are rapidly accumulating intellectual property rights are very effective in extracting the maximum value from preexisting public research, without contributing much to make comparable progress possible in the future. Worse yet, in the current rush to restructure the economy by reducing the public sector to a minimum, the now popular neoliberal policies being put in place are preventing the development of a new generation of basic scientific research.

Taking Stock of Intellectual Property

Fighting against the corporate confiscation of creativity is an uphill battle. Despite its importance, many people probably find intellectual property to be a very abstract, perhaps even forbidding subject, remote from the forefront of public concerns. Indeed, intellectual property rights hardly register in the public consciousness, except when specific issues, such as the Napster controversy, come into popular view.

Earlier, I mentioned that the very term "intellectual property" helps to obscure what is at stake in the ever-expanding grasp of intellectual property rights. Until fairly recently, the use of the expression "intellectual property" was uncommon. People occasionally referred to "literary property," since the private ownership of the author was generally clear-cut. However, the use of the more general expression "intellectual property" was very rare. An opinion in an 1845 circuit court case, *Davoll v. Brown,*[10] mentioned intellectual property. An 1873 Supreme Court decision, *Mitchell v. Tilghman,*[11] quoted a letter using the phrase, which did not appear in a Supreme Court decision again until 1949, in *C. I. R. v. Wodehouse.*[12]

The practice of labeling the monopolistic privileges of patents, copyrights and trademarks as intellectual property conveniently insulated them against criticism. Although a small minority of people may express scorn for intellectuals, for the most part society has a high regard for the intellect. More important, today, property occupies a hallowed position in political discussions in the United States. Combining the two terms as "intellectual property" makes these monopoly rights virtually unassailable.

Attacks on property of any kind have emotional associations in the contemporary United States, which cast those who would whittle away at these monopolies as thieves who would deprive owners of their rightful property. In other words, the monopolists become the upholders of justice, while those who challenge them take on the roles of callous miscreants who must be punished, or at least restrained. The irony is, of course, that the privileges associated with patents, copyrights, and trademarks are monopolistic, and as such are incompatible with a free market.

At first, my critical assessment of intellectual property rights may seem counterintuitive, given the effusion of fanciful rhetoric concerning the so-called New Economy. To be sure, modern technological wonders seem to be the hallmark of contemporary society. Each day the media recounts tales of new medical breakthroughs. In electronics, too, magnificent technological developments sprout up everywhere.

I recognize that the emergence of these impressive technologies coincides with the strengthening of intellectual property rights, but intellectual property rights have done relatively little to further this technology. Instead, much of the credit belongs to a long history of basic science and research that was mostly conducted without the benefit of intellectual property. In fact, virtually all of the great advances in modern technology are the result of research, both in basic science and technological applications, provided by public institutions or public funds. Without that prior groundwork, much of the current boom in technology would have been impossible.

Besides, until recently, intellectual property rights were not very important either within Internet software or in the production of computer hardware. In fact, this absence of strong intellectual property rights explains why fierce competition has kept computer prices falling so rapidly.

I will show how the spread of intellectual property rights, rather than promoting technological progress, threatens to cripple progress in medicine, as well as in science in general. For example, I will show how the pharmaceutical companies have displayed an almost total disinterest in the African market. With a shocking disregard for human life, they actually refuse to produce some drugs that are known to be effective for diseases endemic to Africa. Presumably, the number of people who can afford to pay for these medicines is too small to make production worthwhile.

While individual problems associated with intellectual property, such as the AIDS scandal or the litigation associated with Napster, may engage people, I am certain that we need to see these particular cases as symptoms of a larger crisis of intellectual property rights.

The Choice Ahead

Society has three options regarding the current system of intellectual property. First, we could trust the free market to allocate the fruits of knowledge and information. To do so, I will argue, would prove to be an unmitigated disaster. Although you might not think so from reading the business press, in reality free markets are incompatible with intellectual property. In the absence of secrecy, virtually no business could make a profit by putting time and resources into the development of science or information, since competitors could take advantage of such investments without having to invest anything. Secrecy, however, is inimical to progress. In addition, the dogma of free markets precludes public investment in science and technology as an alternative to private investment.

The second option is to follow the present course by continuing to grant those who supposedly develop intellectual property the exclusive right to its benefits for a set period of time. I do not deny that the promise of gain from intellectual property rights might cause somebody, someplace, sometime to create some new, socially useful technology. But I will make the case that intellectual property rights will shrink, rather than expand, the economic pie. The sweeping scope of intellectual property rights represents a serious threat to both scientific and technological progress.

Such a regime of intellectual property also creates an atmosphere of secrecy, which will certainly inhibit progress. In addition, it leads to an excessive duplication of efforts because companies attempt to work around the intellectual property rights of others. Eventually, it will lead to so much litigation that progress can be brought to a virtual standstill.

The third option is to treat the knowledge and information that society develops as a social good—the property of society as a whole. This option has never been fully attempted, but it seems to be the most promising of the three.

In a sense, the period immediately following World War II was an amalgam of the second and third options. In the United States, researchers working under the auspices of the government or within the university system engaged in a massive but uncoordinated technological enterprise. At the same time, private corporations were able to capture much of the benefits of this research and transform it into private intellectual property.

Presently, public research is withering. Even where government funding is available, conditions attached to the research usually contaminate the outcome. Researchers now must conform more than ever before to the narrow interests of the corporate sector. This corporate pressure threatens the future prospects for society.

Let me repeat for emphasis: while the plethora of new technologies that come on the market almost daily might create the impression that the corporate sector is performing magnificently, in truth, these rapid advances are the fruits of previous scientific efforts, often dating back many decades. Cutting off public support for research and technology and relying on the corporations threatens to stifle the flow of technological progress in the future.

In short, the ongoing elevation of the place of intellectual property will fall quite short of its promise on a number of counts. Nobody can predict the full costs of the course we are now following, but they are certain to be substantial.

CHAPTER ONE

THE ASCENSION OF INTELLECTUAL PROPERTY RIGHTS

The Controversial Origins of Intellectual Property

Proponents of intellectual property rights seem to regard them as the pinnacle of the market economy. This perception seems to parallel a changing vision of the economy. For the last two centuries, the market appeared to be a particular organization of society to create material objects. In recent years, it has been seen more as a means to foster the development of information.

In reality, intellectual property rights are antithetical to the market. In fact, intellectual property rights were "born as a response to market failure," in the words of Robert Merges, a professor of law at the University of California, Berkeley, who specializes in intellectual property.[1] In other words, because the market was not providing the sort of result that the rulers desired, they used the incentive of intellectual property to improve the situation.

Merges was correct in a second sense. Intellectual property becomes most attractive at the very times when crises engulf markets. But first, some background might be useful. The Venetians invented the patent in the fifteenth century. As Italian craftsmen—particularly glass workers—fanned out across Europe, they brought with them the idea of legal protection for invention. In return for bringing their expertise, these craftsmen would enjoy monopolistic privileges.[2] Eventually, patents came to Great Britain under William Cecil (Lord Burghley), chief minister under Elizabeth I, who used patent grants to induce foreign artisans to introduce Continental technologies into England. In effect, then, patents were originally a vehicle for stealing information from others rather than promoting invention. Daniel Defoe, the author of *Robinson Crusoe,* proudly wrote of the British success in this regard:

> It is the kind of proverb attending the character of English men that they are better to improve than to invent, better to advance on the designs plans which other people have laid down, than to form schemes and designs of their own. . . . [Most] of our great advances in arts, in trade, in government, and in almost all the great things, we are now masters of, and in which we so much exceed all of our neighboring nations, are really founded upon the inventions of others.
>
> Even our woolen manufacture itself, with all the admirable improvements made upon it by the English, since it came into their hands, is part of building upon other men's foundations, and improving on the inventions of the Flemings: The wool indeed was English, but the wit was all Flemish; we had the materials, but no more understood the virtue of making them, then the world understood the making of gun power, tho' they had always the sulfur and the salts, which are now the proper ingredients of that dreadful composition.[3]

Notice one difference between these early experiments in intellectual property rights and today's situation. While trade was important, the world then was much more fragmented than today. Intellectual property rights were meant to broaden access to information, by luring foreign expertise into a country. Today, with the economy becoming more globally integrated, intellectual property rights serve to limit access to information.

In one sense, early intellectual property rights took on a modern coloration. By the early seventeenth century, James I of England used the patent as a favor to be dispensed to well-placed courtiers. Under this rubric patents were granted on such enterprises as running alehouses. Parliament responded to this abuse with the Statute of Monopolies of 1624, which forbade all grants of exclusive privileges except those described in its famous Section 6, which described patents for new methods of manufacturing.[4]

Today, rather than courtiers, great corporations who can afford to buy powerful politicians enjoy favored access to intellectual property rights. While the Parliament acted to prevent the king from dispensing such favors, in the contemporary United States both Congress and the executive branch fall over each other to strengthen the intellectual property rights of their favorite patrons.

This cursory history of the early evolution of patents suggests that intellectual property can be used for a multitude of purposes, including efforts to gain advantage over other societies and to provide bounty for favored players as well as to encourage invention. All three motives are at work in the emerging regime of intellectual property rights.

Despite the present acclaim for intellectual property as the source of universal prosperity, critical analysis of it was once widespread. Patents were especially controversial in England, France, Germany, Holland, and Switzerland during the period 1850 to 1875, particularly among those who believed in free trade and laissez faire.[5] In other words, the people who believed in the market most fervently criticized patents as a violation of laissez faire.

According to two distinguished scholars of intellectual property, Fritz Machlup and Edith Penrose, "At the end of the 1860s the cause of patent protection seemed completely lost."[6] But then, the world economy suddenly fell into a period of prolonged crisis, which became known as the Great Depression—at least until the 1930s, when the world experienced another Great Depression.

This Great Depression of the late nineteenth century set off powerful competitive pressures, which ravaged producers. Faith in markets quickly dissolved, lending credence to those who favored intellectual property rights. In the midst of this horrific market failure, arguments that intellectual property represented a violation of free-market principles carried little weight.

Those principled free traders who opposed intellectual property rights as a violation of laissez faire lost ground to the protectionists. In the words of Machlup and Penrose, "The idea of patent protection regained its public appeal when, after the crisis of 1873, protectionists won out over free traders."[7]

As I will show later, political economic conditions were a far more decisive factor than ideological arguments in promoting the rise of intellectual property rights. Conveniently, intellectual property rights allowed major corporations to circumvent the recently enacted Sherman Antitrust Act of 1890. Before this legislation, corporations had routinely ignored the intellectual property of independent inventors. Protection of intellectual property helped to reinforce the powers of giant corporations, which could afford their own research laboratories.

Similarly, in recent decades, as the economy faltered at the end of the 1960s, setting off panic in many corners of the economy, suddenly economic and political leaders turned to intellectual property rights as a means of propping up the economy. United States corporations, which were rapidly losing ground to foreign competitors, turned to intellectual property rights as a way of extracting huge profits.

In conclusion, rather than symbolizing the pinnacle of market success, intellectual property rights are an expression of the failure of the market. Patents and other intellectual property rights come to the fore when markets threaten to self-destruct.

A Dickens of a Problem

While defenders of intellectual property are quick to point out its constitutional origins, they are less inclined to acknowledge that the Constitution was fairly limited in its protection of intellectual property rights. Only those who were either citizens or residents of the United States enjoyed intellectual property rights. For example, publishers in the United States were permitted to publish freely works written by foreigners. This approach made perfectly good sense. Given the underdeveloped state of the American economy, people in the United States were more consumers than producers of intellectual property in the international market.

Popular British authors were outraged by this practice—perhaps none more so than Charles Dickens. During his tour of the United States in 1841, he hectored the audiences at his lectures about the injustice done to authors like him. He reported to a friend:

> I spoke, as you know, of international copyright, at Boston; and I spoke of it again at Hartford. My friends were paralysed with wonder at such audacious daring . . . every man who writes in this country is devoted to the question, and not one of them dares to raise his voice and complain of the atrocious state of the law . . . of all men living I am the greatest loser by it. My blood so boiled as I thought of the monstrous injustice that I felt as if I were twelve feet high when I thrust it down their throats.[8]

Earlier, in *Nicholas Nickleby,* Dickens launched his attack on the failure to recognize copyrights. He mocked a literary gentleman "who had dramatised in his time two hundred and forty-seven novels as fast as they had come out—some of them faster than they had come out." He put the following words in the mouth of his protagonist:

> "Shakespeare dramatised stories which had previously appeared in print, it is true," observed Nicholas, "but he improved on the stories." Violating copyrights was something else: "Now, show me the distinction between such pilfering as this, and picking a man's pocket in the street: unless, indeed, it be, that the legislature has a regard for pocket-handkerchiefs, and leaves men's brains, except when they are knocked out by violence, to take care of themselves."[9]

Later many popular American authors, such as Mark Twain, Walt Whitman, and Louisa May Alcott, sided with Dickens because the Europeans were pirating their work.[10] In 1853 a federal Circuit Court rejected Harriet Beecher Stowe's claim that a German translation of *Uncle Tom's Cabin* infringed her copyright.[11] The absence of royalties for foreign authors also put their works at a competitive disadvantage inside the United States. Since the costs of publishing a Dickens novel included no royalties, while those of a Twain novel did, American writings became relatively more expensive.[12] In 1989, more than two centuries after the creation of its Constitution, the United States finally ratified the Berne Convention of 1886, reflecting the nation's development as a creator as well as a consumer of intellectual property.

One other factor entered into the calculation: during the late nineteenth century, the world was in the grip of the Great Depression of the nineteenth century. As I noted earlier, in the midst of this massive market failure, defenders of free-market principles lost considerable ground to those who would grant monopoly protection to holders of intellectual property rights.

Today, another century has passed. Now, the United States is the largest purveyor of intellectual property in the world. In line with its self-interest, it is coercing much of the rest of the world to accept an unprecedented array of intellectual property claims.

Rock against the Copyright

Like Dickens, music composers also felt abused by those who refused to pay royalties. Just as in Dickens's case, in which the absence of royalty payments helped to popularize European literature in the United States, the state of intellectual property rights had a profound cultural impact.

The story begins at the turn of the century, when the famous composer, Victor Herbert, heard musicians play his music at a restaurant. He sued the owner of the restaurant under a 1909 copyright law. After initially losing in a lower court, Herbert won in the United States Supreme Court in the 1917 decision.[13]

Already in 1914, Herbert had gathered eight composers, publishers' representatives, and lyricists for a meeting that ultimately would result in the establishment of the American Society of Composers, Authors, and Publishers (ASCAP) as their collection agent. In the wake of Herbert's legal triumph three years later, ASCAP expanded its fee-seeking horizons beyond theaters and dance halls to any place where performance for profit occurred.

By 1923, ASCAP turned its attention to broadcasting, selecting WEAF, AT&T's powerful New York outlet, as its test case. Since AT&T was absolutely dependent upon intellectual property, it was ill positioned to oppose ASCAP, and so settled on a one-year license of $500 in payment for all of the ASCAP-licensed music WEAF chose to air. This blanket license arrangement eventually became the industry standard.

> Perceiving themselves to be at ASCAP's mercy, major market station owners formed the National Association of Broadcasters (NAB) in 1923 to do battle with the licensing organization. Station radio concerts were far less appealing without the melodies ASCAP controlled, but the annual license fees, which escalated upward . . . were seen as too high for many stations (few of which had much revenue, let alone profits at this point) to pay. In subsequent years, ASCAP used its near-monopoly position in the music industry to charge broadcasters ever-higher rates.
>
> In 1931 . . . ASCAP boosted its overall fees to stations by 300 percent, charging 5 percent of each outlet's gross income. It then broke off dealings with the NAB and began negotiations with individual broadcasters, offering three-year contracts beginning at 3 percent of net income for the first year, 4 percent for the second, and the full 5 percent only by the third. By 1936 it was demanding five-year licenses.[14]

In 1939, ASCAP raised its fees once again. By the following year, the radio industry had established its own licensing organization, Broadcast Music Incorporated (BMI). On January 1, 1941, the radio stations ceased broadcasting ASCAP music. Instead, they turned to music that was in the public domain, such as Stephen Foster tunes. Surprisingly, the public did not react negatively.

Broadcasters then turned to other sources of music outside of ASCAP's traditional constituency. The intensive broadcasting of Latin American music eventually led to the popularity of the rumba, samba, and tango in the 1940s. Later, BMI recruited composers of blues, country, and gospel music. Much of BMI's success, however, came from rock-and-roll, which became the dominant form of popular music for decades. In the end, then, a considerable amount of popular culture may represent an attempt to circumvent the unduly restrictive powers of intellectual property.

The story of rock-and-roll suggests that intellectual property rights are not absolute. In the case of music, the public was willing to do without much of the music that ASCAP was offering. Unfortunately, substitutes for many of the other materials protected by intellectual property rights are more difficult to come by.

Corporate Resistance to Intellectual Property

The changing nature of corporate research also had a substantial effect on the state of intellectual property rights. Rather than pursuing intellectual property rights, corporations in many industries actively collaborated with one another in an effort to improve existing technologies. For example, in the U.S. iron industry, "if a firm constructed a new plant of novel design and that plant proved to have lower costs than other plants, these facts were made available to other firms in the industry and to potential entrants."[15]

Prior to the crisis of 1873, most large firms were consumers rather than producers of new technology. They depended upon the work of the independent inventors, whom the patent system was supposedly intended to protect. Corporations were not inclined to respect such intellectual property rights, preferring to pay as little as possible or even nothing at all to inventors.

Consider the behavior of the railroads, which dominated the economy of the time. These huge corporations routinely trampled the patent rights of independent inventors. When the inventors did receive compensation, it was limited, because the railroads were so powerful.[16] When companies rode roughshod over the work of independent inventors, sympathetic courts occasionally awarded the owners of these patent rights three times the amount of money that the invention was estimated to have saved the firm. Because the railroads usually prevailed in court, they saved money by paying the infrequent triple-damage award. In the early 1870s, in response to the violation of patent rights for braking systems, the federal courts in Illinois twice affixed damages of several hundred dollars per car for each year of service.[17]

The railroads successfully appealed the judgment. The Supreme Court ruled in their favor in October 1878 in the famous Tanner case. Not only did the Court support the railroads, it even based its ruling on arguments that the railroad industry provided the judges. Justice Bradley confidently wrote, "Like almost all other inventions, that of double brakes came when, in the progress of mechanical improvement, it was needed; and being sought by many minds, it is not wonderful that it was developed in different and independent forms." Expressing a philosophy of technical change in which the railroads and others who employed patented technologies could find great comfort, he continued, "[I]f the advance towards the thing desired is gradual, and proceeds step by step, so that no one can claim the complete whole, then each is entitled only to the specific form of device which he produces."[18]

Once the major corporations began to develop their own research capacities, they conveniently reversed their position on intellectual property. As new producers of intellectual property, they became appreciative of the efforts required to produce inventions. At this point, they expected that these sacred property rights be respected.

The Sherman Antitrust Act of 1890 represented a major milestone in the evolution of intellectual property rights in the United States. Because this law forbade horizontal mergers, many firms turned to in-house research and development as a means of promoting corporate growth.[19] The giant corporations suddenly discovered another potential of their inventions: their intellectual property rights provided a powerful tool to limit competition without violating the provisions of antitrust law. David Mowery and Nathan Rosenberg, two specialists in the economic history of technology, wrote about a slightly later example:

> Patent licensing provided a basis for the participation by General Electric and Du Pont in the international cartels of the interwar chemical and electrical equipment industries. U.S. participants in these international market-sharing agreements took pains to arrange their international agreements as patent licensing schemes, arguing that exclusive license arrangements and restrictions on the commercial exploitation of patents would not run afoul of U.S. antitrust laws.[20]

The next stage in the development of intellectual property came a couple of decades later, when litigation over intellectual property rights began to undermine modern industry—especially industry with crucial military applications. At that point, the federal government stepped in and put a halt to the litigation, requiring all concerned to work together. In effect, the government realized that the morass of litigation would inhibit the development of military technology. As a result, they shelved the existing patent law and forced the companies to cease their disruptive litigation.

I shall return later to the use of patents as a way of circumventing the antitrust laws and to the way in which the government intervened in disputes over intellectual property rights. For now, I will close this section merely by noting that the support for patent rights seems to display an especially slippery tendency toward expediency.

At times when intellectual property rights are convenient for those who wield the most power in society, supporters of intellectual property rights pretend that the protection afforded by patents offers a rare combination of efficiency and morality. At other times, when intellectual property rights inconvenience powerful interests, they are dismissed out of hand. This hypocrisy continues today, as we shall see, since many of the staunchest corporate defenders of intellectual property rights stand frequently accused—often with good cause—of violating the intellectual property rights of others.

The Coronation of the Corporation

Modern commentators on the patent system frequently quote with approval the words of Abraham Lincoln, inscribed outside of the Patent and Trademark Office, that patents add "the fuel of interest to the fire of genius." Lincoln himself received Patent No. 6,469 for a device to lift boats over shoals on May 22, 1849.

Today, however, individual genius does not figure very prominently in the ownership of intellectual property. Instead, the bulk of intellectual property belongs to the great corporations.

The conversion of intellectual property rights from the domain of the individual genius to the balance sheets of corporations parallels a more general transformation in modern capitalist economies. Originally, a corporate charter was a special privilege. Petitioners had to convince a legislative body that their project would serve the public interest before they could obtain a corporate charter. Often, petitioners would engage in subterfuge, promising to do one thing while intending to do another, but the fact remained that the legislative body had the power to annul the charter if it chose to do so.

In the course of the nineteenth century, special interests agitated for corporate charters to be a right rather than a privilege. Louis Hacker, along with Charles and Mary Beard, contended that the railroad lawyers who wrote the Fourteenth Amendment to the United States Constitution carefully crafted it to lay the groundwork for an even broader claim; namely, that corporations should have the same rights as individuals.[21]

Even though at least one of the people who framed the amendment was quite clear in describing his role, many modern scholars prefer to attribute the amendment to a humanitarian effort to help the freed slaves. They never explain the reason for this sudden concern for the well-being of the former slaves. Regardless of the motives of the framers, the courts soon began to use the amendment to justify granting corporations all the rights that individuals enjoy under the Constitution.

Individuals, however, lacking the power of corporations, soon found their rights substantially diminished relative to the rights of corporations. In the case of intellectual property, the protection of patents drifted from the rights of individuals and eventually became the property of corporations. Intellectual property, of course, is but a single example in a larger transformation of corporate privileges.

An individual stockholder of a large corporation enjoys all of the benefits of intellectual property rights resulting from the scientific work of employees of the corporation. The logic behind the stockholder's rights relies on a curious twist of legal reasoning. This peculiar quirk of the law treats the corporation as an individual who can take credit for the creation. The shareholder, in turn, as the owner of a part of that fictitious individual, can profit from all the protections of intellectual property rights.

This interpretation of intellectual property rights is a relatively modern innovation. Originally, the patent law did not allow corporations to win patents as inventors. James Boyle, a professor at Yale Law School, has traced the curious process by which the concept of the individual inventor that formed the foundation of the original patent law gradually mutated to include corporations as inventors.[22] This transformation of intellectual property rights came just in time to rescue the giant corporations from the antitrust laws.

Antitrust and Intellectual Property

About the same time that the courts were conferring the rights of private individuals upon the corporations, Congress passed the Sherman Antitrust Act of 1890, which restricted certain anticompetitive behaviors. Although the Sherman Antitrust Act prohibited corporations from acting together in an anticompetitive fashion, it did nothing to prevent the same corporations from merging. As a result, the late nineteenth and early twentieth centuries witnessed a merger wave of unprecedented proportions. The giant corporations that emerged from this process wielded tremendous political influence. In addition, the giant corporations, especially those applying chemistry, began to create large industrial research laboratories, in an effort to catch up with German corporations, which, incidentally, lacked patent protection. About that same time, the courts first began to recognize corporate patents.[23]

Although the Sherman Antitrust Act prohibited the firms from behaving collusively, the corporations soon realized that the patent system gave them an opportunity to accomplish what might otherwise have been prohibited.[24] They collected patents as rapidly as possible, even on processes that they had already used for some time, to gain control over their markets, despite the prohibitions of the Sherman Act.[25] One AT&T patent lawyer explained how his company used the protection of intellectual property to make life difficult for the company's competitors:

> It appears to me that the policy of bringing suit for infringement on apparatus patents is an excellent one because it keeps the concerns which attempt opposition in a nervous and excited condition since they never know where the next attack may be made, and since it keeps them all the time changing their machines and causes them ultimately, in order that they may not be sued, to adopt inefficient forms of apparatus.[26]

This behavior had nothing to do with the ostensible purpose of the patent law, which was supposed to encourage technological research. Instead, the corporate sector as a whole entered into a massive drive to shore up profits by way of patents. At the same time, individual inventors became less and less significant within the patent system. According to David Noble, a historian who has developed a deep understanding of the corporate control of science and technology:

> Although the first patent pool, among manufacturers of sewing-machine parts, was established as early as 1856, it was not until the end of the century that corporations clearly became the dominant factor in patent exploitation. In 1885 twelve percent of patents were issued to corporations; by 1950 "at least three-fourths of patents [were] assigned to corporations."[27]

Noble cited one critic who complained in 1949, "today it would be more correct to say that the patent system adds another instrument of control to the well-stocked arsenal of monopoly interests . . . [because] it is the corporations, not their scientists, that are the beneficiaries of patent privileges."[28] Noble also described the efforts of Edwin J. Prindle, a mechanical engineer and patent lawyer, who "in numerous articles . . . outlined the means of securing patent monopolies to bypass the antitrust laws [and] methods of securing patents from inventors, and employee-inventors."[29]

Senator Joseph Christopher O'Mahoney of Wyoming, presiding over an enquiry conducted by the Temporary National Economic Committee of President Roosevelt (1938-39), summarized testimony concerning the manufacture of bottles and other glass containers in the United States in stark terms:

> Here is an industry . . . where the method of employing patents has resulted in a sort of private N. R. A. [National Recovery Administration]. This control is employed to adjust and allocate production . . . prices are stabilized through production control and . . . are further stabilized through the practice of producers to follow the prices of the largest producer. . . . Through the refusal

> to grant licenses, persons desiring to enter the industry have not been allowed to do so.[30]

Similarly, in *Standard Oil Co. (Ind.) v. United States* (1931),[31] the Court ruled:

> If combining patent owners effectively dominate an industry, the power to fix and maintain royalties is tantamount to the power to fix prices. Where domination exists, a pooling of competing process patents, or an exchange of licenses for the purpose of curtailing the manufacture and supply of an unpatented product, is beyond the privileges conferred by the patents and constitutes a violation of the Sherman Act.[32]

What these earlier commentators said about the patent system—that much of the effort in collecting patents has little to do with the protection of vital research—holds true today more than ever. Many companies merely intend to accumulate a strategic set of patents to block potential competitors from advancing. In the words of Thomas G. Field, a professor of law at the Franklin Pierce Law Center in New Hampshire, "A patent is like a hammer—you could use it to build a house or to kill someone."[33]

Between the two world wars, both the Patent Office and the courts, angered by the anticompetitive use of patents, were considerably less willing to grant questionable patents.[34] Since then the pendulum has swung the other way. Unfortunately, whatever constructive uses patents may have are giving way to even more lethal objectives. For example, as I will discuss later, companies obtain patent rights for widely known phenomena to demand payments from other companies that are merely applying relatively common knowledge.

Corporate vs. Individual Creativity

Summing up, corporations routinely ran roughshod over patent laws until the economic climate changed. Once antitrust laws began to threaten them, they suddenly became advocates of patent protection. The patent laws, originally designed to inspire the genius of individual inventors, began to benefit the industrial laboratories of the major corporations.

This transformation of the locus of economic activity from the individual to the corporation should have sparked the interest of economists. After all, economic theory revolves around the analysis of how individuals respond to the marketplace. However, economists did not seem to embrace this subject.

Almost a century after the Fourteenth Amendment, this tension between the corporation and the individual found a brief mention in the literature of economics, beginning in a very abstract book, entitled *Theoretical Welfare Economics,* published in 1957. The author, Jan Graaf, seemingly innocuously remarked, almost in passing, "In itself a firm possesses no knowledge. That which is available to it belongs to the men associated with it."[35]

Presumably, nobody even noticed this passage for several decades, although eventually it inadvertently stirred up an ideological hornet's nest, at least as far as Sidney Winter was concerned. In 1982, Winter first rebuked Graaf for his seemingly unobjectionable assertion, proclaiming, "it is the firms, not the people who work for the firms, that know how to make gasoline, automobiles, and computers."[36] A couple of pages later, he tempered his position, acknowledging, "I conclude that there is much merit both in the view that individuals are repositories of productive knowledge and in the view that business firms and other organizations are such repositories."[37] A few years later, Winter repeated his earlier critique of Graaf, but this time without any qualification.[38]

Why did Winter keep repeating his critique of Graaf? Kenneth Arrow, who without reference to Winter once observed:

> What is the role of the firm in standard economic theory? It is a locus of knowledge, as embodied in a production possibility set. But where is this knowledge located and in what sense is it characteristic of the firm? Some of the knowledge that is most important is largely embodied in individuals.[39]

Here we come to the rub. Arrow continued:

> The embodiment of knowledge in workers contradicts the standard theoretical meaning of a firm. In the neoclassical model, workers are not part of the firm. They are inputs purchased on the market, like raw materials or capital goods. Yet they carry the firm's information base, even though not permanently attached to the firm. Defining the firm as a locus of productive knowledge leads to a dilemma; what knowledge is peculiar to a firm.[40]

Like Arrow, Winter realized that Graaf's perspective threatened the structure of conventional economic theory. More important, Graaf's approach called into question the entire distribution of income: If individuals are the repositories of knowledge, how do the corporations manage to acquire so much of the fruits of such knowledge?

Later, Winter achieved a measure of fame, along with a coauthor, Richard Nelson, for creating a vision of the firm as a suprahuman organism that embodies an immense quantity of knowledge that would be far beyond the capacity of any individual to manage.[41] In doing so, they attempted to rescue economic theory from its anachronistic foundation, in which everything centers on the private individual. Rather than evaluate the success of their theoretical venture, I will first turn to the subject of the perversion of intellectual property and then address the question of how well the corporation manages science in a world of intellectual property rights.

The Trademark as Power

While corporations were consolidating their patent rights in the early twentieth century, they were perfecting a relatively new use of intellectual property. At the time, some companies realized that they could profit magnificently if they could manage to blunt the force of competition by finding a way to make their product seem special. This challenge was especially daunting for companies that produced commodities without any distinguishing features. The idea that a company could prevent another from using its trademark to sell similar goods first appeared in American law in the middle third of the nineteenth century. Even then, the protection afforded by trademarks was somewhat limited (Fisher 1999).

Trademarks soon became an important element of market power. For example, Quaker Oats created the concept of a breakfast cereal and then set out to convince customers that somehow its oats were superior to those that people had traditionally bought in bulk at the grocery store.[42] Similarly, people had previously bought biscuits in open containers. A combination of biscuit makers, known as the National Biscuit Company and later as Nabisco, began to package their biscuits so that they could market them as an identifiable brand.[43]

Unlike patents, which, as I will show later, can directly undermine technological progress, the proliferation of successful brands has little direct effect on technology. Even so, once brands become established, competition weakens. As a result, the incentive to innovate falls by the wayside.

Brands mostly effect the distribution of income between the owners of the brand and the rest of society. The ability of sellers of branded goods to charge an extra markup on their goods has a powerful effect on the distribution of income. For example, shortly after the branding of food products began, the farmer's share of the average dollar spent for food began to fall and has continued to do so ever since. Farmers attribute this trend to the power of those who process, distribute, and sell food. However, even for foods that have not changed much over the decades, such as bread, the farmer's share has suffered the same erosion.

This trend is not confined to food. Cheap sneakers have given way to expensive running shoes and cross trainers. These shoes, which cost only a few dollars to produce, frequently sell well in excess of $100, creating billionaires in the process. In effect, the purchaser of a branded product pays twofold: once for the product itself and once for the intellectual property embodied in the logo. The owners of the more prominent brands soon realized that the profit potential from their brands exceeded that of the commodity itself. People, for some reason or other, are willing to pay more for the McDonald's experience or the Nike swoosh than for the food or the sneaker itself.[44]

This book will not pay much attention to the effect of the brand names and ubiquitous logos that pollute the environment. Even though the cynical and manipulative application of modern technologies to create needs and dependencies has important economic effects, the analysis of this subject relies more on a cultural critique that others are better suited to pursue. Naomi Klein's *No Space, No Choice, No Jobs, No Logo: Taking Aim at the Brand Bullies*[45] does an excellent job in that regard. I will only mention here that this phenomenon of branding greatly strengthens the grip of the transnational corporations, especially those based in the United States. People around the world seem to be willing to pay a premium for these branded products, ensuring that the poor will continue to pay tribute to the rich.

Continuing Controversies about Intellectual Property

Although the defenders of intellectual property rights achieved dominance with the onset of the Great Depression of the late nineteenth century, some within the libertarian tradition continued to protest that the monopoly powers granted to holders of intellectual property were inconsistent with their idealized version of the market. For example, Friedrich A. Hayek, probably the greatest conservative economist of the twentieth century and a tireless advocate of the virtues of free markets, maintained:

> The problem of the prevention of monopoly and the prevention of competition is raised much more acutely in certain other fields to which the concept of property has been extended only in recent times. I am thinking here of the extension of the concept of property to such rights and privileges as patents for inventions, copyright, trade-marks, and the like. It seems to me beyond doubt that in these fields a slavish application of the concept of property as it has been developed for material things has done a great deal to foster the growth of monopoly and that here drastic reforms may be required if competition is to

be made to work. In the field of industrial patents in particular we shall have seriously to examine whether the award of a monopoly privilege is really the most appropriate and effective form of reward for the kind of risk-bearing which investment in scientific research involves.

Patents, in particular, are specially interesting from our point of view because they provide so clear an illustration of how it is necessary in all such instances not to apply a ready-made formula but to go back to the rationale of the market system and to decide for each class what the precise rights are to be which the government ought to protect. This is a task at least as much for economists as for lawyers. Perhaps it is not a waste of your time if I illustrate what I have in mind by quoting a rather well-known decision in which an American judge argued that "as to the suggestion that competitors were excluded from the use of the patent we answer that such exclusion may be said to have been the very essence of the right conferred by the patent" and adds "as it is the privilege of any owner of property to use it or not to use it without any question of motive." It is this last statement which seems to me to be significant for the way in which a mechanical extension of the property concept by lawyers has done so much to create undesirable and harmful privilege.[46]

Similarly, Lionel Robbins, the conservative economist responsible for bringing Hayek to England, was not particularly supportive of the patent system:

We may leave undiscussed the question whether if there were no patent rights of any kind there would be an inadequate flow of invention. . . . But it is really quite ridiculous to suggest that any modification in the present law would necessarily have this consequence, that the thing is naturally sacrosanct. The patent system in its present form is a highly artificial creation emerging from a process of legislation in which the role of pressure groups and muddled thinking has been unfortunately only too prominent; and no convincing argument has yet been put forward to show that the abolition of the present system and the substitution of a "license of right" system whereby, after a very short period, everyone might use a patent on paying a license fee to the inventor, would be unjust to inventors or diminish the flow of invention. But certainly it would sweep away the whole network of monopolistic practice which rests on the present system.[47]

If markets worked as perfectly as Hayek believed, then the case for intellectual property rights would have to be dismissed out of hand. In part, the problem is, as we shall see, that free markets in ideas and information cannot work. Recall Robert Merges's observations that intellectual property rights reflect a failure of markets.

Unfortunately, as this book will explain, the solution to this market failure proposed by proponents of intellectual property rights does more harm than good. Few economists have taken note of the problems with the present system of intellectual property. One partial exception is Milton Friedman, the doyen of contemporary conservative economists, who once pointed out:

> There are costs involved [in the patent system]. For one thing, there are many "inventions," that are not patentable. . . . Insofar as the same kind of ability is required for the one kind of invention as for the other, the existence of patents tends to divert activity to patentable inventions. For another, trivial patents, or patents that would be of dubious legality if contested in court, are often used as a device for maintaining private collusive arrangements that would otherwise be more difficult or impossible to maintain.[48]

Since Friedman penned this brief note, the problems with intellectual property have become rampant. When Friedman was writing his comment, the patent system had been relatively stable for a long time. Most people had, or at least thought they had, a fair idea of what could or could not be patented. A few economists thought about tinkering with the optimal length of patent protection, but patent protection in itself seemed to be a self-evident way of doing things.[49]

Within a few years of Friedman's observation, the great postwar economic boom began to fizzle out. As the economy threatened to stumble into another crisis, perhaps the most alarming symptom was the balance of trade. Whereas the United States had been exporting massive amounts of goods and services during the postwar boom, by the end of the 1960s the balance of trade began to post deficits, which have continued to balloon ever since.

Leading business and political leaders thought that strengthening the system of intellectual property could help to correct some of the serious structural problems the United States was facing. In the decades that followed, intellectual property rights expanded at an unprecedented rate. Today, life forms, ways of doing business, and virtually everything else seems to be patentable. To my knowledge, nobody has really given a great deal of thought to how these new trends will affect the nature of our economy. I will attempt such an analysis in the present text.

The Newfound Rights of Intellectual Property

So how powerful is this monopoly protection provided by intellectual property rights? The answer varies with the legal and economic climate. The late Edwin

Mansfield was perhaps the leading researcher in the economics of intellectual property. According to Mansfield, when the owners of patent rights earned large profits, other firms managed to find ways to maneuver around the patent in a relatively short time. In one study covering a sample of patents in the chemical, drug, electronics, and machinery industries, Mansfield and his co-workers found that competitors managed to imitate 60 percent of all the patented inventions within four years of their introduction.[50]

Another survey published a few years later more or less replicated Mansfield's results. The authors found that companies must invest time and money to circumvent patents, but they typically succeed. They estimated that "patents raise imitation costs by about forty percentage points for both major and typical new drugs, but about thirty percentage points for major new chemical products, and by twenty-five percentage points for typical chemical products."[51] Richard Nelson, one of the authors of the more recent study, as well as the coauthor of the earlier mentioned book written with Sidney Winter, later concluded on the basis of that survey:

> While I want to emphasize that patents play a much smaller role in enabling innovators to reap returns under modern capitalism than commonly believed, there are certain industries where patent protection is important, perhaps essential, for innovation incentive. Our questionnaire revealed two groups of industries of this sort. One consists of industries where chemical composition is a central aspect of design: pharmaceuticals, industrial organic chemicals, plastic materials, synthetic fibers, glass. The other consists of industries producing products that one might call devices: air and gas compressors, scientific instruments, power-driven hand tools, and so on.[52]

Mansfield and his co-workers, as the authors of the study that Nelson summarized,[53] however, worked in an environment in which intellectual property rights were considerably weaker than they are today. Sadly, nobody has followed up these studies to see how the strengthening of intellectual property rights has affected the ability to imitate patentable products.

In all likelihood, even though the pace of technological change has intensified and imitating patentable products is easier, a new survey would probably find that patents today are even more effective in excluding would-be imitators. Pharmaceutical companies are still able to produce knock-off drugs. In 1997, according to the Boston Consulting Group, which provides advice to the industry, 42 of the 100 top-selling medicines were me-too drugs, representing 47 percent of top drug sales.[54] For example, Merck supposedly produced a product similar to Celebrex, a new $1.4 billion arthritis medicine from Monsanto Co., within five months.[55]

Such a quick response, however, is exceptional. Even when a company can create an imitation drug, years can pass before it can bring the product to market. Meanwhile, the original drug can bring in hundreds of millions of dollars each year.

In other cases, a drug can remain dominant by dint of advertising and promotion, even when comparable substitutes are available. For example, the rights to Claritin remain so valuable that when Schering-Plough Pharmaceuticals faced the prospect of the expiration of its patent on the drug in 2002 with nothing promising in its research pipeline, it went to Congress for help. The legislators seemed ready to oblige by passing what Representative Henry Waxman derided as the "Claritin Monopoly Relief Act" of 1999 that would have extended the patent an additional three years, until a wave of protests arose.

Then, in the summer of 2000, there was an attempt to slip language into a military-appropriations bill at the last minute to extend the patent.[56] One study from the University of Minnesota estimated that the extension would cost consumers $7 billion.[57] Although other drugs are equally effective, they are not quite the same formulation. As a result, the enormous advertising effort for the drug can hold off the competition. If another company had been able to reverse-engineer the drug successfully, Schering-Plough would have little to gain by extending the patent.

The reason for the urgency of Schering-Plough's patent-extension efforts is obvious. Once a patent expires, other firms begin to sell generic substitutes. Within a month of the patent's loss, a typical drug will lose half of its sales; within a year, sales usually drop 80 percent. No wonder. Many patients pay $2.50 or more per dose. Generic drug makers, which still make a healthy profit, report they could produce an equivalent pill for 50 cents.[58]

Pharmaceutical companies use whatever legal maneuvers they can to delay the introduction of generic drugs. As a last resort, they sometimes pay millions of dollars to generic drug makers to make them agree not to market a generic drug.[59] While the two corporations profit from the deal, consumers must pay much more for their medications.

The Political Economy of Intellectual Property

The logic of political economy explains the obliging attitude of the government. Recall that the longstanding surplus in the balance of trade of the United States turned negative. As manufactured goods poured into the United States in the late 1960s, government leaders agonized about finding ways to increase exports. Strengthening intellectual property rights seemed to be an ideal strategy for promoting exports from the United States.

The most powerful multinational corporations in the United States took advantage of this climate and energetically lobbied for stronger intellectual property rights.[60] A valuable journalistic study of the subject described the change in the perception of intellectual property:

> [A]s a flood of imports washed away millions of domestic manufacturing jobs, attitudes toward patents and their role in the economic equation began to change. The interests of industry and labor coalesced in the search for viable weapons in the fight against foreign competition. The election of Ronald Reagan further shifted the mood toward protection of intellectual property. The major philosophical argument against patent protection—that it was inherently monopolistic—was no longer politically or, even more to the point, economically correct in an era of increasing trade competition. The policy of using antitrust laws against companies that refused to license their patent technologies was reversed by the Justice Department.[61]

Despite this philosophical change, the U.S. government's first reaction to growing imports was to coerce individual countries to restrain their exports. In the words of Paul David, an economic historian who specializes in matters concerning science and technology:

> [D]uring the 1980s, the U.S. government responded to the concerns of American producers—especially chemical, pharmaceutical, electronic and information technology industries—by trying to reverse the trend of the preceding two decades. Acting with some encouragement from other industrially advanced countries, the United States pursued a direct, unilateral course of action. It did not make any major effort to negotiate agreements within the framework of the Paris Convention for the Protection of Industrial Property (patents and trademarks); the Berne Convention for the Protection of Literary and Artistic Works (copyrights) or other international conventions, nor did it offer some quid pro quo to other developing nations that would agree to such conventions. Instead, by threatening within the context of bilateral trade negotiations to impose sanctions on developing and newly industrialized nations whose retaliatory leverage was quite limited, the United States achieved considerable leverage in convincing foreign governments to acquiesce to its position on the treatment of various forms of intellectual property.[62]

Then, the government reversed course, but not without prodding. In the words of Edmund J. Pratt, Chairman Emeritus of Pfizer:

> In 1983, Pfizer joined with other corporations such as Merck, Johnson & Johnson, Bristol-Myers, IBM, Hewlett Packard, General Motors, General Electric, Rockwell International, Du Pont, Monsanto, and Warner Communications to form the Intellectual Property Committee to advocate intellectual property protection. The committee helped convince U.S. officials that we should take a tough stance on intellectual property issues, and that led to trade-related intellectual property rights being included on the GATT agenda when negotiations began in Punta del Este, Uruguay, in 1986.[63]

Then, in 1986, six months before the Punta del Este meeting launching the Uruguay Round of the General Agreement on Tariffs and Trade, the chief executive officers of 12 major corporations belonging to the committee met again. The result was profound: "The IPC [Intellectual Property Committee], in conjunction with its counterparts in Europe and Japan, crafted a proposal based on existing industrialized country laws and presented its proposals to the GATT Secretariat. By 1994, the IPC had achieved its goal in the Trade Related Aspects of Intellectual Property (TRIPs) accord of the Uruguay trade round. . . . In effect, twelve corporations made public law for the world."[64]

Domestic law also changed to enhance intellectual property rights. The history of the semiconductor industry exemplifies this trend toward changing the legal structure to help firms gain a competitive advantage from intellectual property rights rather than encouraging them to develop an edge in productive capabilities.

As late as 1981, Roger S. Borovoy, vice-president and chief counsel for Intel Corporation, declared, "In the electronics industry, patents are of no value whatsoever in spurring research and development."[65] A recent study published by the Philadelphia branch of the Federal Reserve System describes the dramatic transition that came soon after Mr. Borovoy's evaluation of the importance of the patent system to his industry:

> Within the U.S. semiconductor industry, reverse-engineering was a well-established practice. But by the late 1970s, American firms objected to similar behavior by Japanese firms when they began to increase their market share in the more standardized products, such as computer memory chips. The level of competition eventually became so intense that, by the mid 1980s, most American companies abandoned these segments entirely.
>
> When it became clear they could no longer dominate Japanese firms on the basis of production technology alone, American firms attempted to consolidate their comparative advantage in research and development. To do this, they would have to find ways of reducing their competitors' ability to

> reverse-engineer their products. To that end, American companies began to lobby Congress to increase intellectual property protection for their semiconductor designs. In 1984, Congress created a new form of intellectual property right, called mask rights, especially tailored to address the needs articulated by the industry.[66]

During this period, both Texas Instruments and National Semiconductor were tottering on the verge of bankruptcy. Irving Rappaport, former vice-president and associate general counsel for intellectual property at National Semiconductor, recalled:

> I'm not exaggerating when I tell you that National Semiconductor was only weeks away from bankruptcy in late 1990. . . . All the papers had been signed before it was decided to continue the business and give licensing a more aggressive push. And without a doubt, patent fees bought us valuable time in which to complete our restructuring process. For a while there, in fact, three-quarters of our revenues came from patent licenses.[67]

Texas Instruments struck first. Typically license fees ran about 1 percent of revenues. In 1987, Texas Instruments raised its royalties on chips to 5 percent.[68] The company filed a suit against one Korean and eight Japanese semiconductor companies, accusing them of infringing semiconductor patents. The settlements yielded the company more than $600 million in payments, according to a 1990 report. The company became so aggressive in seeking royalties that by 1992 it earned $391 million in royalties, compared to an operating income of only $274 million.[69]

Other companies are even more successful. For example, IBM's annual report announced that the firm had earned more than $1.5 billion in income in 2000 from its intellectual property portfolio.[70]

In effect, these companies are beginning to transform the semiconductor industry from a manufacturing industry to a service industry, just as the postindustrial utopians would have them do it. According to one industry insider, James Koford of LSI Logic, "Silicon Valley and Route 128 are worlds of intellectual property, not capital equipment and production. Most of the employees of U.S. high technology live in southeast Asia."[71] Some, like George Gilder, applaud this arrangement, arguing that these companies will rationally maximize their profits by specializing in the design of computer chips.[72]

This new legal framework allowed some of the most important high-technology sectors of the economy to become a perverse kind of service sector.

Rather than directly providing services, they merely demanded payment for their intellectual property.

Not only did the administrative arm of government become more sympathetic to the rights of intellectual property, the judicial system, with vigorous support from the legislative side, also moved to embrace intellectual property rights more warmly. In 1982, the Federal Courts Improvement Act established the Court of Appeals for the Federal Circuit as a central patent appeals court to streamline the litigation process and create a judicial body for appellate review of the patent infringement decisions of the lower courts.[73]

This court has been very sympathetic to the claims of patent holders. Previously, the percentage of court decisions that upheld patents had been decreasing from roughly 40 percent in the 1950s to less than 30 percent in the 1970s, suggesting a possible tightening of standards.[74] The new court, in a stunning reversal, now upholds patents 80 percent of the time.[75]

Prior to 1986, federal courts frequently decided the validity of a patent based on a preponderance of the evidence standard; that is, which side presented more convincing evidence. In 1986, the Federal Circuit ruled that a patent should be presumed to be valid until proven otherwise by clear and convincing evidence, a more difficult standard to satisfy.[76]

This trend toward upholding a larger share of patents comes at a time when the scope of patents has broadened at an unprecedented rate. For example, in recent years, the patent system has begun granting claims for computer software, life forms, and even business practices. When moving into such previously uncharted terrain, a growing number of questionable proposals would be expected to come up for consideration. Nonetheless, the proportion of patents approved has continued to increase.

Each year, the government of the United States moves to strengthen intellectual property rights still further. Patents now last 20 years instead of 17, despite the fact that high technology typically becomes obsolete in less time than ever before. In addition, a new law recently has transformed many types of copyright infringement into criminal offenses.

At the time of this writing, furthering the powers of domestic holders of intellectual property in the international marketplace is one of, if not the highest priority of the government of the United States. The creation of the World Trade Organization (WTO) in 1995 symbolized this expanding role of intellectual property. The forerunner of the WTO, the General Agreement on Tariffs and Trade, had the mission of liberalizing trade in goods. The WTO, in contrast, vigorously enforces trade in intellectual property.

The Corporate Stakes in Intellectual Property

From the standpoint of exports alone, this emphasis on intellectual property proved highly successful. In 1947, intellectual property comprised just under 10 percent of all U.S. exports. By 1986, the figure had grown to more than 37 percent. By the early 1990s, the best estimate was that intellectual property accounted for well over 50 percent of exports from the United States.[77] In 1999, U.S. exports in the form of royalties and licensing revenue alone exceeded $37 billion—topping aircraft exports, at $29 billion, and telecommunications equipment. Moreover, the trade surplus in intellectual property—the exports minus imports—is running at about $25 billion annually, and growing.[78] As already noted, IBM alone enjoyed worldwide licensing revenues that exceeded $1.5 billion, according to its annual report for 2000.[79] These figures exclude payments for physical goods, such as computer chips, which also embody intellectual property.

Other companies are even more successful. For example, IBM's 2000 annual report announced that the firm had earned more than $1.5 billion in income in 2000 from its intellectual property portfolio.

Intellectual property goes well beyond high technology. Jack Valenti, chairman and chief executive officer of the Motion Picture Association of America, proudly announced to the House Committee on Ways and Means, Subcommittee on Trade:

> The U.S. copyright-based industries, which include the motion picture, sound recording, computer software, and book industries, were America's number one export sector in 1996. These industries together achieved foreign sales and exports of $60.18 billion, surpassing every other export sector, including automotive, agriculture and aircraft. The U.S. filmed entertainment industry alone earned about $12 billion in foreign revenues in 1997, 40 percent of the total revenues earned by the U.S. film industry.[80]

The revenues from intellectual property are certain to grow. The wealthy nations, led by the United States, make poor countries adopt intellectual property laws similar to those of the United States as a condition of joining the World Trade Organization. Henry Nau, who later became a senior staff member for international economic affairs on Reagan's National Security Council, accurately observed that the logic of this new regime increased global interdependence and made it possible for the West to employ their superior scientific and technical expertise to impose "a more subtle and total form of imperialism than was possible in any previous period of history."[81]

While the impressive revenues from intellectual property rights provide lavish benefits to the largest transnational corporations, the enforcement of intellectual property rights has ominous implications for the rest of society that I will discuss throughout this book.

Recall that the impetus for imposing this new regime of intellectual property rights was to correct the imbalance between exports and imports in the United States. While exports have grown impressively, imports have soared even more because of the declining emphasis on manufacturing in the United States, leaving the U.S. economy with the largest balance of trade deficit ever recorded. In the absence of intellectual property revenues, however, the balance of trade would probably be even worse.

Defending and Reforming Intellectual Property

The defenders of intellectual property will naturally take issue with my skepticism regarding intellectual property rights. They base their advocacy on the proposition that profit incentives are absolutely necessary to induce people or corporations to innovate. Their exaggerated claims do not lend credibility to their position. For example, Richard Parsons, president of AOL Time Warner, the world's largest copyright owner,[82] offered this preposterous warning:

> I think this is a very profound moment historically. . . . It's about an assault on everything that constitutes the cultural expression of our society. If we fail to protect and preserve our intellectual property system, the culture will atrophy. And corporations won't be the only ones hurt. Artists will have no incentive to create. Worst-case scenario: The country will end up in a sort of cultural Dark Ages.[83]

Of course, this concern for the larger interest of society is rank hypocrisy. In fact, the same companies that become so upset when Napster distributes their music offer music themselves on their Web sites without paying royalties to the publishers—even when the publisher is a division of the same company.[84]

Despite their solicitous concern for creative artists when justifying their intellectual properties, people such as Richard Parsons must know that the vast majority of cultural workers earn little or nothing from their creative endeavors. Virtually all of the proceeds that the corporate sector passes on to the creative workers goes to a tiny fraction of them. For example, in 1999, just 88 recordings—only .03 percent of the compact discs on the market—accounted for a quarter of all record sales.[85] Jon King, a performer in the group Gang of Four, observed, "the vast majority of musicians don't actually get paid, they in fact have day jobs."[86]

While the corporations protest when a consumer distributes copies of their intellectual property to others, the corporate publishers routinely resell copies of stories that they publish to the electronic media without compensating the writers. Freelance writers are presently suing, claiming that the corporations are violating their intellectual property rights. I will discuss this suit in more detail later.

Richard Parsons's defense of intellectual property can draw more credible support from the so-called Schumpeterian hypothesis, named after Joseph Schumpeter, certainly one of the most important economists of the twentieth century. Schumpeter suggested that monopoly, in general, increases the potential for research and development, because only the large firm can muster the resources necessary to create modern technological advances.[87] Skeptics counter that the lack of competition associated with monopoly can reduce the pressure to improve productivity, causing research and development to slacken off.[88]

Schumpeter was writing in the context of large, monopolistic, industrial firms, such as the leading producers of steel or automobiles. His idea was that an aluminum monopoly would devote enormous resources to research and development because of the threat to its markets from a steel monopoly, since factories can substitute one metal for another.

The monopoly rights conveyed to owners of intellectual property are not comparable to those of industrial monopolies, especially because these intellectual properties frequently do not compete with one another. A researcher looking for the cure to a disease associated with a particular gene will not shop around to find a cheaper gene to cure the disease. I can think of some cases where the Schumpeterian hypothesis might hold, but, on the whole, I believe that the current regime of intellectual property significantly threatens the future development of technology.

Perhaps the most fanciful defense of intellectual property rights comes from Ayn Rand, a novelist and philosopher who had great influence on the libertarian movement; including the current chair of the Federal Reserve Board, Alan Greenspan. In a book that consisted of a collection of her articles, except for two essays by Greenspan, Rand proclaimed in all seriousness: "Patents and copyrights are the legal implementation of the base of all property rights: a man's right to the product of his mind . . . patents are the heart and core of property rights, and once they are destroyed, the destruction of all other rights will follow automatically, as a brief postscript."[89]

Of course, today intellectual property rights do not usually protect "a man's right," but rather that of the corporation that employs him or her. Even so, Rand reminds us of the perennial challenge: In the absence of markets, who would create the supply of information that underlies intellectual property?

Admittedly, in a market society, many people would be unwilling to devote much time and energy to producing a product that was free for the taking, unless a supplementary return, such as advertising, subsidized the product. Nonetheless, some exceptional people have been willing to do what is supposedly irrational—foregoing the profits from a scientific invention. On April 12, 1955, amidst the immense public excitement created by the trials of the Salk polio vaccine the late TV commentator, Edward R. Murrow, asked Jonas Salk, who would control the new pharmaceutical. The doctor replied, "Well, the people, I would say. There is no patent. Could you patent the sun?" Similarly, many people have contributed to various open-source software projects for no remuneration at all.

Such occasional examples do not constitute proof that enough people would have the same spirit of generosity as a Jonas Salk. What alternatives are there to the present system of intellectual property?

Michael Kremer, a very clever economist at Harvard University, has suggested an interesting procedure to reward the production of intellectual property, while avoiding some of the distortions associated with treating intellectual property as a monopoly.[90] He proposed that patents be put up for an auction. The winning bidders would be allowed to keep some of the patents, but the state would purchase the majority for the price of the winning bid, plus a premium. The state would then put its patents in the public domain.

Kremer's method seems superior to the situation as it exists today. Even so, it represents a modest step, at best. Kremer would allow some of the patents to remain as private monopolies. More important, this procedure would still privatize the rewards from scientific research, leading to the secrecy that seriously threatens the entire scientific process. Later, I will go into detail about the harm that secrecy does.

Mickey Mouse Economics

Intellectual property rights extend to matters that are totally unrelated to science and technology. Presently, the policy of the U.S. government places great weight on the protection of such intellectual property, despite the fact that the main beneficiaries are a handful of global corporations.

The Copyright Extension Act of 1998 is symbolic of the increasing grasp of intellectual property. The Constitution originally provided for 14 years of monopoly rights. By 1998, Congress had already extended the copyright to 50 years after the death of the author. The 1998 law provides holders of copyrights an additional 20 years of protection in addition to the previous 50 years after the death of the author. Supposedly, this revision was urgent because Disney's rights to Mickey Mouse were due to expire in 2003. Without this law, the poor mouse would have had to endure life within the public domain rather than within the secure confines of the Disney empire's holdings of intellectual property.

Now, recall that the stated purpose of the copyright laws is to "promote the progress of science and useful arts." I have not yet been able to discover the link between extension of the Mickey Mouse copyright and the promotion of science and the useful arts, but I am sure that wiser people than me are well aware of the merits of this legislation.

Even so, the arts did relatively well without the full protection of intellectual property rights. Authors, such as Charles Dickens, prospered despite the absence of royalties from the United States. This tradition extends back even further. As Dickens himself noted, Shakespeare built his plays from other people's books. He lifted *Romeo and Juliet* from *The Tragicall Historye of Romeus and Juliet,* a 1562 translation of a tale by Matteo Bandello, an Italian writer, soldier, and monk. *Julius Caesar* came out of Sir Thomas North's 1579 translation of Plutarch's *Lives of Nobel Grecians and Romans.* Lawrence Lessig, a Harvard law professor, noted: "Extending a dead author's copyright won't encourage him to write another book." Lessig wondered, "Would Shakespeare have written *Romeo and Juliet* if he had to get permission from Bandello and pay royalties?"[91]

The government's efforts go well beyond shoring up the legal rights of holders of this kind of intellectual property. The full weight of its power is brought to bear against all evildoers who would dare to create knockoffs of a Disney cartoon or a Nike swoosh. In the words of Thomas Friedman, perhaps the most enthusiastic proponent of globalism at the *New York Times:*

> The hidden hand of the market will never work without a hidden fist—McDonald's cannot flourish without McDonnell Douglas, the designer of the F-15. And the hidden fist that keeps the world safe for Silicon Valley's technologies is called the United States Army, Air Force, Navy and Marine Corps. . . . Without America on duty, there will be no America Online.[92]

Lest the skeptical reader dismiss Friedman's clever phrasing as nothing more than a rhetorical flourish, consider the words of William Cohen, the secretary of defense in the Clinton administration. In February 1999, upon his arrival in Seattle—a city that a few months later became a symbol of resistance to the policies that he was sent to advocate—to speak to the employees of Microsoft, the secretary told reporters, "I will point out that the prosperity that companies like Microsoft now enjoy could not occur without having the strong military that we have."[93]

Thomas Friedman and William Cohen expressed what is probably the central thrust of the foreign policy of the government of the United States. I suspect that the response of the U.S. government to Saddam Hussein would have been far more severe had he engaged in counterfeiting intellectual property rather than merely invading a neighboring country.

Associating the protection of intellectual property with national defense might seem far-fetched, but, I am convinced that much of the defense of intellectual property is equally preposterous. Consider one more piece of evidence. Recent history demonstrates fairly conclusively that the government of the United States signals a high priority for any policy initiative by associating it with national security. In the case of intellectual property, the Clinton administration raised the stakes to an even higher level by invoking the threat of terrorism. An unsophisticated reader might wonder how the counterfeiting of corporate logos and the like might threaten the life and limb of citizens of the most powerful nation on the face of the earth. Apparently, the government let it be known that it is seriously concerned that people counterfeiting T-shirts and ripping off logos might actually be raising funds for terrorists.[94] Naturally, the government has no choice but to use every power available to it to prevent such dangerous activities.

The terrorism threat goes well beyond T-shirts. On May Day 2001, Assistant U.S. Attorney Daniel Alter warned a federal appeals court hearing arguments in *Universal Studios v. Reimerdes et al.* about the dangers surrounding the DeCSS utility, a tiny program that allows computers with the Linux operating system to read digital video disks. The Motion Picture Association of America sued to take the program off a Web site. In support of the industry, Mr. Alter likened DeCSS to other tools that terrorists might use, such as "software programs that shut down navigational programs in airplanes or smoke detectors in hotels." He warned: "That software creates a very real possibility of harm. That is precisely what is at stake here."[95]

Mickey Mouse Economists

Considering the economic importance of patents, a surprisingly small number of economists have shown any interest in them. Writing in the early 1950s, Fritz Machlup, who perhaps more than any other economist devoted his career to analyzing the implications of an economy in which the collection and use of information play an important role, observed:

> Judging from the share which the subject of patents has had in the literary output of economists of the last fifty years, and from the share which economists have had in the literature on the subject of patents, one may say that economists have virtually relinquished the field. Patent lawyers were probably glad to see them go; some said as much with disarming frankness.[96]

To get an idea of what Machlup meant, I went to JSTOR, an Internet site that stores articles from academic journals.[97] The economics collection includes around 40,000 articles, some dating from the late nineteenth century. Only a handful of the articles contain the words "patent" or "patents" in the title. Prior to an article that Machlup published in 1950, the collection includes a mere dozen items concerning patents.

The economists who followed Machlup were a little more active. Even so, a mere 12 articles appeared during the 1960s, followed by only 5 in the 1970s. Then in the 1980s, 22 articles appeared. This modest spurt continued between 1990 and 1994, the present cutoff point for the collection. Of the 6 articles published during this interval, 3 appeared as part of a single symposium.

Relatively few of these articles even attempt to acknowledge any doubt about the efficacy of the patent system. Some merely use patents as indicators of the value of information or intellectual property. Others look at the frequency of patent renewals. Only a couple of the modern articles raise any questions at all about the overall rationality of the patent system.

Articles with the term "intellectual property" in the title are even scarcer. A scant 6 exist in JSTOR, all from the 1990s. One of these also has the word "patent" in the title. Three of these articles appear in the symposium mentioned earlier. So here is the supposed engine of the modern economy, yet, as in Machlup's time, it barely registered in the economics literature—at least until very recently.

I suspect that most economists would regard patents somewhat favorably because of the widespread belief that patents encourage innovation. Yet, in the words of two of the more thoughtful students of the subject, "some of the most innovative industries today—software, computers, and semiconductors—have historically had weak patent protection and have experienced rapid imitation of their products."[98]

Despite the generally favorable climate of economic opinion regarding patents among modern neoclassical economists, this book will make clear that patents and other intellectual property rights are creating serious obstacles and confusion that will ultimately stand in the way of technical progress. The way that the current regime of intellectual property rights is unfolding leads me to expect that the ultimate result of the continued expansion of intellectual property rights will be an economy dominated by secrecy and litigation. Technical progress will be stunted, while intellectual property rights distort the economy in ways that disadvantage all but the fortunate few.

CHAPTER TWO

THE PERVERSIONS OF INTELLECTUAL PROPERTY

The Primitive Accumulation of Intellectual Property

Intellectual property rights cover a wide variety of activities, ranging from protecting Nike Corporation from producers of counterfeit tennis shoes to granting exclusive rights to an author. Supposedly, these intellectual property rights serve to encourage creators, whether authors or inventors, to exert themselves in a manner that will benefit society as a whole.

The logic of intellectual property is not nearly as straightforward as it might seem at first glance. For example, within the context of the existing system of intellectual property rights, a band of traditional people in the Amazon who discover important medicinal properties of a plant supposedly do not deserve the protection of intellectual property rights since no individual agent can claim credit for this knowledge.[1]

The idea of a debt to an indigenous band in the tropics is not an idle fantasy. Many modern medicines originated in the tropics. According to Mark Plotkin, an ethnobotonist:

> A quarter of all prescription drugs sold in the United States have plant chemicals as active ingredients. About half of those drugs contain compounds from temperate plants, while the other half have chemicals from tropical species. According to one recent study, the value of medicines derived from tropical plants—that is, the amount consumers in the United States spend on them—is more than $6 billion a year.[2]

Early communities never treated biological or genetic information as intellectual property rights. Instead, they protected their "economic" interests through either secrecy or mandates. For example, the coffee plant, native to Ethiopia, was surreptitiously taken to Yemen, where it became an important item of commerce. According to Mark Pendergrast, a journalist who devoted an entire volume to the subject of coffee:

> The Turks jealously guarded their monopoly over the trees in cultivation in Yemen. They prohibited any fertile berries from leaving the country without having first been steeped in boiling water or partially roasted to prevent germination.
>
> Inevitably, these precautions were circumvented. Some time during the 1600s a Moslem pilgrim named Baba Budan smuggled seven seeds out by taping them to his stomach and successfully cultivated them in southern India, in the mountains of Mysore. In 1616 the Dutch, who dominated the world's shipping trade, managed to transport a tree to Holland from Aden. From its offspring the Dutch began growing coffee in Ceylon in 1658. In 1699, another Dutchman transplanted trees from Malabar to Java, followed by cultivation in Sumatra, Celebes, Timor, Bali and other islands in the East Indies.[3]

Such efforts to keep their biological heritage to themselves represented an early analog to the present-day enforcement of intellectual property rights. Later, British plant explorers, including Charles Darwin, went to great lengths to conceal their clandestine theft of plant material, which allowed Britain to prosper mightily at the expense of countries whose biological heritage Britain tapped. For example, rubber plants found in Brazil were bred in the British Royal Botanic Gardens to be made suitable for Malaysian plantations.[4]

Although individuals or even small communities have no formal rights to intellectual property, some countries are now claiming compensation for the use of genetic information garnered from living things found within their borders. These countries base their demands on the Convention on Biological Diversity, which grew out of the Earth Summit in Rio de Janeiro in 1992. According to this convention, which the government of the United States has arrogantly refused to ratify, nations have sovereignty over their genetic resources and are entitled to "fair and equitable sharing of the benefits."

For example, Abbott Laboratories is presently developing a painkiller modeled on the active chemical in the secretions from the skin of a frog. Generations of tribesmen in the Amazon rain forest have long used this material to make poison blow darts. This drug seems as effective as morphine but without damaging side effects. Inconveniently, Ecuador and many other developing countries are demanding compensation in exchange for permits that allow scientists to collect biological samples.

When the governments of the tropics request compensation from the pharmaceutical companies that want to take advantage of the rich plant life of the area, the corporations express outrage, complaining that such demands will impede scientific development. A recent *New York Times* article echoed the pharmaceutical companies' displeasure regarding the stance of the countries that are demanding compensation: "Once considered the common heritage of mankind, wild animals, plants and crops were taken without asking or freely exchanged."[5]

Four points are worth noting here: First, the article neglected to mention that this scientific development will be unlikely to provide many benefits for the people of that part of the world, especially the most disadvantaged people. Second, the argument about the common heritage applies just as well to scientific research in general. Abbott Laboratories is not likely to "freely" exchange its discovery; the company wants to collect frogs so that it can make a large profit by patenting a product with exclusive market rights. Third, as I will show later, intellectual property claims by North American corporations do far more to bottle up the common heritage of science than the comparatively modest demands of the South American governments. Finally, while the article cited above is correct in asserting that historically genetic material flowed across borders, this movement was not exactly free, as the earlier reference to Darwin suggests.

Although the *New York Times* may scoff at the audacity of poor countries in standing in the way of science, in truth the scientific accomplishments of these seemingly primitive societies were outstanding, insofar as knowledge of biology is concerned. According to the late Edgar Anderson, who was associated with Missouri Botanical Garden of St. Louis, during the last 5,000 years modern society has not domesticated a single plant that primitive cultures had not already used. He pointed out that traditional cultures had already managed to discover all five natural sources of caffeine: coffee, tea, the cola plant, cacao, yerba mate and its relatives.[6]

The emerging business of growing Japanese wasabi roots using modern agricultural technology illustrates the tensions between secrecy—the traditional equivalent of intellectual property—and its modern variant. In the 1970s, Roy Carver III, a Southern California real estate developer, accepted a cotton farm as payment for one of his office buildings. He wanted to farm, but preferred something more lucrative than cotton. In 1989, he assembled a group of consultants to explore crops with better profit potential.

When Carver and his associates traveled to Japan in 1991 to begin research on growing wasabi, they were told by Japanese scientists that wasabi could not be grown anywhere but Japan. The plant requires a particular combination of environmental conditions: cold, clear, running water; a cool climate; not too much sunshine, preferably at the bottom of a deep valley, supposedly unique to a specific location in Japan.

Undaunted, Carver set out to obtain books on wasabi growing, but he was informed that the books were unavailable outside of Japan. As in the case of coffee, the wasabi growers wanted to protect their traditional livelihood, although they lacked any legal basis for their case.

Finally, Carver had a Japanese woman purchase the books. He then photocopied them and had different translators work on separate parts of the books so that no individual would be familiar with the whole process. Eventually, Carver and his team designed elaborate greenhouses capable of growing the wasabi root under control conditions. He now sells wasabi for almost $60 per pound.[7]

Having appropriated knowledge of wasabi growing through a ruse, Carver now goes to great pains to protect his own knowledge from others. He surrounds his ranch with an eight-foot security fence and employs guards so that other people will not be easily able to replicate what he has done.

Later, I will indicate that the corporate world, just like the primitive world, relies heavily on secrecy. This tactic is closing off an ever-increasing swath of scientific activity. In contrast, the traditional scientific method relies on openness and communication as standard operating procedures. A common theme running throughout this book will be the desirability of the scientific method relative to the sort of secrecy and monopolization of knowledge and information characteristic of the corporate-dominated system of intellectual property.

Radio Days and the Easing of Intellectual Property Rights

When intellectual property rights become too complex, wasteful litigation over ownership can consume the attention of industry. Nobody can move forward because of the thicket of intellectual property rights that stand in the way of progress. In this perverse environment, technological progress can come to a standstill.

Even so, I have no doubts that the initial creation of intellectual property rights will have the effect of stimulating activity on the part of those who seek profits. The important question remains: What happens as the conflicting intellectual property claims accumulate?

Immediately after the creation of the intellectual property rights, researchers are still free to draw upon a valuable pool of nonproprietary research. This combination of new opportunity, together with the accessibility of freely available prior research, acts as a short-run stimulant to technological progress. Over the long run, however, the privatization of scientific knowledge will severely retard technological development as litigation consumes an increasing amount of effort. In addition, a growing cloud of secrecy, together with the exploding expense of paying holders of intellectual property rights, will hinder technological progress.

The prospect of intellectual property rights inhibiting research and development stands in sharp contrast to their prevailing justification on the grounds that they effectively promote technological progress. The vibrant pace of contemporary industrial progress seems to suggest that one might comfortably put such worries aside. After all, even in the software and biotechnology industries, where the application of intellectual property rights has created so much controversy, technological advances seem to have proceeded unabated.

The experience of the radio industry in the early twentieth century indicates that the idea that industrial progress could cease because of the threat of intellectual property rights is not merely a theoretical possibility. Instead, it suggests that the world of strong intellectual property rights might not be hospitable to technological progress.

Even though intellectual property rights were far weaker then than they are today, the radio industry faced the real danger that progress would become impossible because so many companies claimed intellectual property rights over the basic elements of the central technologies of the industry.

First some background: Radio was on the verge of becoming the cutting-edge, high-tech industry of the day. Before the age of broadcasting, the U.S. Navy was the primary overseer of the radio industry since ship-to-shore communication was the major use for the technology. At the time, a burst of new inventions rapidly expanded the technological capability of radio, but the Navy was still primarily concerned with the potential of maintaining communications among its ships at sea.

A complex pattern of conflicting patent rights was threatening to forestall the future development of the industry. Major Edwin Armstrong, a key inventor of early radio technology, testified before the Federal Trade Commission, "It was absolutely impossible to manufacture any kind of workable apparatus without using practically all of the inventions which were then known."[8] General Electric, Westinghouse, American Marconi, AT&T, and even United Fruit all held vital patents.

With the outbreak of war, the government could no longer countenance the individualistic nature of these conflicting claims to intellectual property rights. According to Ruth Schwartz Cowan:

> [Already in] 1915, the government ordered the end to the dozens of patent suits that were holding up the manufacturing of various electronic components. A patent pool was created for the duration of the conflict, allowing any company with adequate facilities to manufacture anything that the government needed.[9]

In effect, a welter of conflicting patent claims created what some writers called the tragedy of the anticommons, where "too many owners hold rights in previous discoveries that constitute obstacles to future research."[10] The military was unwilling to trust the fate of such a vital technology to the prevailing process of development within the regime of intellectual property rights.

In addition, the Navy did not see any reason for inventors to earn steep royalties. Instead, when they wanted to purchase a particular apparatus, they would encourage competitors of the holder of the patent to make copies to sell to the Navy at a cheaper price, disregarding patent rights altogether.[11]

The Benefits of Superseding Patents

In 1915, General Electric developed the Alexanderson alternator, a major advance in radio technology. Around March 1919, the British Marconi Company sent its representatives to the United States to negotiate for the exclusive right to the Alexanderson machine and its accessories.[12] Rear Admiral William H. G. Bullard, director of communications of the Navy, and Commander Stanford C. Hooper, of the Bureau of Engineering of the Navy Department, also attended the meeting. Bullard stated that if General Electric sold these devices to Marconi, foreign interests would have a monopoly in worldwide communications.[13]

Both the Navy and independent radio manufacturers agreed that the conflicting patent claims retarded the manufacture and operation of commercial radio systems (Federal Trade Commission 1923, p. 24). On January 3, 1920, Captain A. J. Hepburn, acting chief of the Navy's Bureau of Engineering, sent a letter to AT&T, Western Electric, and General Electric. He began by referring to "numerous recent conferences" held in connection with the radio patent situation and stated the bureau's belief that all interests would best be served by "some agreement between the several holders of pertinent patents" whereby the market could be freely supplied with tubes. Adequate radio communication, he insisted, was a matter of life and death for people at sea.[14]

In one of the conferences to which Hepburn referred, the Navy had urged the founding of a single entity for the combined radio patents of General Electric,

Westinghouse, and American Marconi, later jointed by AT&T and United Fruit, that would best serve national security.[15] Soon after this 1919 meeting, General Electric ceased negotiations with British Marconi and began to work with the Navy to form an American radio company, which eventually became the Radio Corporation of America.[16] By 1923, a government report concluded, "The Radio Corporation has entered into agreements with the various companies which own or control practically all patents covering radio devices considered of importance to the art. The number of patents involved approximates 2,000."[17]

Noobar Danielian, in his extensive study of AT&T, concluded:

> The result of the intensive development and acquisition of patents in the radio held by the competitive interests was a stalemate created by mutual patent interferences. This was appeased during the World War, when the United States Government guaranteed to protect the various manufacturers against all infringement suits. As a result, during this period, vacuum tubes were manufactured without regard to patents, by Westinghouse, General Electric, Western Electric, the American Marconi company, De Forest Radio, and others. Upon the termination of the war, however, A.T.&T. and the electrical manufacturers were at loggerheads. They found themselves fighting each other to occupy a terrain which was not marked by distinct boundary lines. Ownership of the various patents pertaining to vacuum tubes and circuits by different concerns prevented the manufacture of an improved tube for radio use. The Federal Trade Commission in 1923 enumerated a brief list of the important patents required in the construction and operation of vacuum tubes, which were in conflict.[18]

The Navy had another justifiable concern about the negotiations with British Marconi. Earlier, the British seemed to have taken advantage of another monopoly by intercepting the telegraph traffic that flowed through the underwater cables that connected Europe and the United States. In 1921, a Senate subcommittee wanted to know if the British also altered or failed to deliver sensitive communications. Questioned closely on the subject, Newcomb Carlton, president of Western Union, refused to answer.[19]

Once World War I was over, the litigation recommenced. Between 1900 and 1941, a total of 1,567 infringement suits entangled 684 different radio patents.[20] These patent suits extracted a heavy price in terms of technological development. Reviewing the history of the British radio industry, one writer observed with a notable understatement that radio manufacturers in Britain wasted "a lot of ingenuity" during the 1920s devising circuit arrangements that reduced the royalties that would otherwise have to be paid to British Marconi.[21] Although radio tube technology advanced in the process, this progress would have been much greater had researchers not directed so much energy to working around existing patents.

Another historian of the radio industry similarly concluded: "In a significant sense, then, science has been compromised to the extent that research funds and researchers have been sacrificed to the essentially unproductive work needed to gain or maintain monopoly position, and pursuit of patents has been at the heart of this process."[22]

The costs of these patent disputes are incalculable. Fortunately, one of the most vibrant centers of radio technology was in the San Francisco Bay Area, especially Palo Alto. This activity represented the beginnings of the phenomenon that caused this region to become known as Silicon Valley. This region had one advantage. RCA's monopolistic practices did not affect these early West Coast electronics firms as much as they did their East Coast counterparts because they were far enough away from RCA and its lawyers or small enough or working in separate niches.[23]

Frederick Terman, professor of electrical engineering, dean, provost and vice president of Stanford, and the person typically credited with "inventing" Silicon Valley, discussed this burst of activity in a 1978 interview:

> I think they were "every man for himself" much more back [east]. . . . [East Coast] manufacturers would never cooperate (on standards for vacuum tubes), partly because of the patent situation. RCA dominated the patents and you couldn't leave RCA out, and if RCA was brought in, it wanted to boss everything. The group out here was involved in military production, instruments, and specialized stuff, where RCA patents weren't such a dominating feature. RCA wasn't trying to build a monopoly in the instrumentation business, for example.[24]

The openness and informality of the region, as contrasted with the legalistic quest for intellectual property, became the hallmark of Silicon Valley, while the more restrictive work style continued in the East Coast electronics firms. I will return to this thought in the next chapter.

Aircraft and Semiconductors

The history of the aircraft industry differed from that of the radio industry in one major respect. In the radio industry, a number of competing patents bogged down the industry. In contrast, in the aircraft industry, the Wright brothers held a single basic patent, which gave them substantial control over the industry. Consider the verdict of Robert Merges and Richard Nelson. The former is the law professor mentioned earlier as a specialist in intellectual property law. The latter was discussed earlier as Sidney Winter's coauthor. He has also been at the forefront of economists attempting to understand the role of technology within the firm. In their words:

> Other creative people and companies, however, wanted to enter the aircraft design and manufacture business. They had their own ideas about how to advance the design of aircraft, and they strongly resisted being blocked by the Wright patent. The early attempts by the Wright brothers and Glen Curtiss, who was the most prominent such potential competitor, to reach an agreement came to naught. Litigation followed. . . . The situation was so serious that, during World War I, at the insistence of the secretary of the navy, an arrangement was worked out to enable automatic cross-licensing.[25]

As a result of the government intervention, the industry was able to manufacture a plane in far less time than would otherwise have been required.[26] In addition, Merges and Nelson concluded: "There is good reason to believe that the Wright patent significantly held back the pace of aircraft development in the United States by absorbing the energies and diverting the efforts of people."[27]

While the patent struggles were destructive, the patent pool that the government helped to organize also proved to be an impediment to the development of the industry. Taking advantage of their situation, the major aircraft corporations used their pooled patents to undermine potential competitors. By 1972, the government finally charged the industry with antitrust violations.[28]

In the absence of military pressure, both the radio industry and the aircraft industry might have been left to languish in a morass of litigation over patent rights. Instead, partially freed from the confining constraints of intellectual property rights, radio technology in particular forged ahead into areas that neither the industry nor the military had foreseen.

A similar process occurred later in the semiconductor industry. Texas Instruments won a patent for its invention of the integrated circuit in 1964. In the same year, Fairchild Instruments obtained a patent on the planar process for producing them cheaply. Turning again to Merges and Nelson:

> The Department of Defense, which for some time had provided the lion's share of the markets for semiconductors, traditionally has tried to avoid being at the

> mercy of a firm or small number of firms with key patents. Through a variety of maneuvers it has encouraged general cross-licensing, and this set the tone for deliberations to resolve the problem created by the two patents. . . . It is hard to argue that this has slowed down the development of integrated circuits.[29]

Merges and Nelson suggest that the Defense Department may have also had a hand in ensuring the absence of a patent for the basic design for the computer. Summing up, they conclude: "there are many cases in which technical advance has been very rapid under regime of intellectual property rights were weak or not strongly enforced."[30]

If intellectual property rights are so efficient, why does the government step in during wartime emergencies or when intellectual property claims threaten to bottle up the development of a military technology? Why not just allow holders of intellectual property to pursue their conflicting claims in the courts?

In the cases of radios and semiconductors, the military intervened to prevent the major corporations from devouring each other in wasteful litigation. In the case of the aircraft industry, the monopolistic side of intellectual property prevented vigorous technological development. Once the cooperating aircraft firms began to use their collective patent rights to stifle competition, defeating the initial effort to weaken the Wright brothers patents, the government had to re-enter the industry.

In short, once intellectual property rights were weakened, each of the three industries—radio, aircraft, and semiconductors—became more dynamic. In each case, the military acted to promote the development of a crucial industry in the name of national defense.

The current discussions about intellectual property typically display less understanding of the stakes than the military seems to have exercised. Instead, the idea of intellectual property remains shrouded in a haze of rhetoric. The modern student of the subject is led to believe that the promotion of ever-stronger intellectual property rights is the key to the development of high technology.

The semiconductor industry probably benefited from the weakening of intellectual property rights in another respect. Intel, the industry giant, seems to have profited from access to the research of another company:

> Intel, the microprocessor manufacturer, was started, not in a garage or basement as many other Silicon Valley start-ups, but when Robert Noyce, the General Manager of Fairchild Semiconductor, and Gordon Moore, its head of Research and Development walked out of Fairchild and set up their own firm,

Integrated Electronics. Shortly before their departure, a scientist in Moore's department had discovered the "silicon-gate" technique to produce semiconductor memory devices. This became an important part of Intel's proposed product line. As a former employee of both companies put it, "Intel was founded to steal the silicon gate process from Fairchild." Clearly, of all Fairchild's employees, Noyce and Moore had the greatest access to Fairchild's inventions, and at the very least, took a lot of knowledge and, equally important, employees with them to the start-up. Thus, Intel hit the ground running, and is now one of the most profitable firms while Fairchild Semiconductor is virtually a footnote in business history. The circumstances of Intel's founding are not an exception. Bhide reports that 71 percent of the firms included in the Inc 500 (a list of young, fast growing firms) were founded by people who replicated or modified an idea encountered in their previous employment.[31]

The Bizarre World of Intellectual Property Rights

Some examples of intellectual property claims would be humorous if the courts did not take them so seriously. Lawyers are now suggesting that athletes should patent the way they shoot a basket or catch a pass.[32]

You may recall the earlier discussion about the zeal of the American Society of Composers, Authors, and Publishers (ASCAP) in pursuing radio stations for royalties. Ever on the lookout for more royalties, it even was about to sue the Girl Scouts for singing songs such as "Row, Row, Row Your Boat" while sitting around campfires until adverse publicity caused it to relent.[33] On the same day that the Girl Scout article appeared, a *Wall Street Journal* article reported that the National Basketball Association was engaged in a suit against America Online over the transmission of game scores and statistics from NBA games in progress.[34] In another case, someone, in all seriousness, patented the correct way of lifting a box.[35]

In one famous case, a patient found that his doctor had patented genetic material from the patient's own body without informing him. The patient sued for compensation, but the courts upheld the doctor's rights to the intellectual property encoded in the patient's genes.[36]

Absurd claims to informational property rights have been expanding by leaps and bounds. People have successfully convinced the Patent and Trademark Office to grant property rights for everything from colors to even a specific number.[37] The Patent and Trademark Office even registered the "frowny" emoticon :-(as a trademark of Despair.com.

Ralph Lauren won a victory in an appeals court in 2000, when his lawyers forced a magazine, begun in 1975 as the official publication of the U.S. Polo Association, to change its name because Lauren claimed the word "Polo" as intellectual property.[38] In a similar case, when educators at the Australian Institute of Management listed a 20-year-old course, "Effective Negotiation Skills," on the organization's Web site, a United States training group, Karrass, told the institute to take the course description off the site because Karrass has a U.S. trademark over the words "effective negotiating," "advanced effective negotiating," and "effective sales negotiating."[39]

One critic of the patent system succeeded in winning a patent for Kirchoff's law, a scientific principle first developed in 1845, proving that the electric current flowing into a function equals the current flowing out.[40] If an individual critic of the patent system is able to manipulate the Patent and Trademark Office into registering such ridiculous claims, think of how much profit-maximizing corporations with enormous resources available for research and legal expenses are able to stake out as private property.

To illustrate this point, Richard Stallman, winner of a MacArthur "genius" award, challenged Bruce Lehman, then head of U.S. Patent and Trademark Office, at a contentious meeting. Stallman produced a voluminous, unwieldy printout of a computer program he had written earlier with several colleagues. He explained that the program was currently in use on more than a million computers, including those of the U.S. Air Force and major companies, such as Intel and Motorola. "Just a few lines of code can be enough to infringe a patent, and this compiler has ten thousand pages," Stallman said, gesturing to the document. "How many patents does it infringe? I don't know. Nobody does. Perhaps you can read the code and tell me?" he challenged Mr. Lehman.[41]

Each morning, as the director of the Patent and Trademark Office, arrives at his Crystal City, Virginia office, he walks past a framed poster bearing the motto "Our Patent Mission: To Help Our Customers Get Patents."[42] In line with its "business mission," the agency pays its examiners sparingly. Greg Aharonian, publisher of the *Internet Patent News,* commented "You really have to be a patriot to work at the Patent Office."[43]

This particular business is unusual in one respect: its customers want a minimum of service. In other words, all applicants prefer a minimum of scrutiny for their applications. The agency seems to be more than happy to comply. It allows its examiners no more than 80 hours to complete a case from start to finish.[44] The average patent takes considerably less than the maximum allowable time. Supposedly, the average time spent on a patent is a mere 8 hours.[45]

In 1997 the Patent and Trademark Office received 237,045 submissions, which took 4,300 staff years, or an average of less than a person-week to process a submission according to the Patent and Trademark Office Web site. Obviously, some proposed patents are ridiculous and can be dismissed quickly. Even so, an alarming number of silly patents still survive the process and win approval.

The modest salaries given to examiners at the agency further suggest management's entrepreneurial spirit rather than a serious concern with thorough scrutiny of patent claims. Although many of the examiners have advanced degrees in highly technical fields, in 1999 the entry-level job descriptions for patent examiners listed salaries ranging from $20,588 to roughly $60,000. After 10 or 15 years, an examiner who has reached "Primary Examiner" status may earn $72,000 to $80,000, substantially less than they might earn in the private sector.[46] Most of the examiners are lawyers. Salaries of associates at East Coast law firms commonly exceed $140,000.[47]

Michael Kirk, executive director of the American Intellectual Property Law Association in Alexandria, Virginia, and a former assistant patent commissioner, recalled, "I was at the patent office in the 1980s when the first wave of biotech patents hit, and we had to hire examiners to qualify them. . . . We called in some industry folks and asked, 'How much would we have to pay these people?' They came up with a figure higher than the salary of the secretary of commerce."[48]

A former patent office employee told a writer for *Science,* the journal of the American Association for the Advancement of Science, "They are desperate and they're hiring like crazy." Even a spokesman for the agency, Richard Maulsby, admits, "It is very difficult for us to do all this hiring and to maintain quality."[49]

Examiners have a strong financial incentive to approve patents. A case does not count as part of an examiner's output until it is closed. If an examiner denies a patent application, the applicant can appeal, dragging out the process and lowering the agency's estimate of the examiner's productivity. One exasperated examiner wrote, "Hey, management pays you for good patents or bad, right? In fact, why should you fight with management? Why reject?"[50]

To make matters worse, junior examiners complain that they receive very little effective training.[51] Turnover among examiners is high. By the time that they acquire much experience, many examiners leave for the private sector, often to companies where they can use their knowledge of the agency to help their employers navigate the patent process more successfully.[52]

The agency seems to do whatever it can to rush patents along. A patent office employee reported, "When I started several years ago, we were told 'When in doubt, reject.' But now, it's 'When in doubt, issue the patent.'"[53]

Beginning in 1991, Congress ceased providing the Patent and Trademark Office with any tax revenues. Instead, the agency was expected to fund itself from the fees paid by those who applied for patents, who then became seen as customers. Then Congress went even further, demanding that the agency be financially self-sufficient. In 1998, this madness peaked. In that year, the agency spent $126 million on examiners' salaries and $50 million on rent, while Congress diverted approximately $100 million to the federal coffers.[54] Given the frequent fiascos coming out of the patenting process, Congress had little choice but to ease off its demands.

Congress's initial transformation of the Patent and Trademark Office was part of a larger process of attempting to make the government work more like a business. Presumably, this policy also made its customers happier than would an alternative policy of allocating more funds for greater scrutiny of patent applications.

In fact, many powerful interests hoped that the agency would become still more like a business. In April 1997, the House of Representatives even passed H.R. 400, which contained provisions to change the Patent and Trademark Office from a Department of Commerce agency to a government corporation.

Although the Patent and Trademark Office proudly reports on its output as if it were just another business serving its customers, it is not just another business with the mission to satisfy customers. Instead, its purpose is to act as an important arbiter of the development of science and technology. Today, the agency is perhaps the most important referee regarding the distribution of intellectual property rights. In any case, its real customer is society as a whole.

Considering the awesome responsibilities of an examiner, economizing on their salaries and hurrying them in their decisions seems to be foolish, to say the least. A single mistake can throw an industry into turmoil, leading to millions of dollars in litigation costs. Granting an inappropriate patent can sidetrack valuable research efforts. Because investors often interpret a patent as a seal of approval from the government, a sloppy patent system can lead investors to make mistakes. This misinformation can waste resources that could be put to better use.

More important, the examiners are making decisions that will have a significant influence on the overall distribution of wealth and income within the U.S. economy and even globally, because of the hegemonic power of the United States in shaping the status of intellectual property in the rest of the world.

The arbitrary nature of the patent system harms some businesses while it rewards others. Theoretically, companies injured by an unjustified patent can challenge the legitimacy of the decision, but to do so can cost millions of dollars. The costs of challenges go beyond these direct economic costs, because firms necessarily shift resources from research to litigation. Fred Warshofsky claims: "In some fields like biotechnology, for example, legal briefs outweigh scientific papers by orders of magnitude, and lawyers are as eagerly sought as Ph.D.'s."[55]

Often, just paying royalties to the holders of dubious patents is cheaper than attempting to get claims dismissed in the courts. Besides, the courts are no more prepared to understand the underlying issues than the Patent and Trademark Office is. In the process, intellectual property rights divert even more funds from pure research.

In other cases, firms do not even bother to go through the motions of creating intellectual property. Instead, as I will discuss in the next section, they specialize in buying up defunct firms that might have acquired valuable patent rights without having been aware of their importance. The law comfortably embraces such absurdities. Perhaps the final word in this section, then, should go to Marx—not Karl, but Groucho. In 1948, the Marx Brothers were planning to make a movie called *A Night in Casablanca.* Warner Brothers, a corporation founded by distant relatives of mine, threatened suit, saying that it owned the rights to the name. Groucho Marx retorted in a zany letter that brilliantly captures the absurdity of the world of intellectual property rights:

> Dear Warner Brothers,
> Apparently there is more than one way of conquering a city and holding it as your own. For example, up to the time that we contemplated making this picture, I had no idea that the city of Casablanca belonged exclusively to Warner Brothers. However, it was only a few days after our announcement appeared that we received your long, ominous legal document warning us not to use the name Casablanca.
>
> You claim you own Casablanca and that no one else can use that name without your permission. What about "Warner Brothers"? Do you own that, too? You probably have the right to use the name Warner, but what about Brothers? Professionally, we were brothers long before you were. We were touring the sticks as the Marx Brothers when Vitaphone was still a gleam in the inventor's eye, and even before us there had been other brothers—the Smith Brothers; the Brothers Karamazov; Dan Brothers, an outfielder with Detroit; and "Brother, Can You Spare a Dime?" (This was originally "Brothers, Can You Spare a Dime?" but this was spreading a dime pretty thin, so they threw out one brother and gave all the money to the other one and whittled it down to, "Brother, Can You Spare a Dime?").[56]

The Abuse of the Patent System

The patent system as it stands is an open invitation to abuse. Symbolic of the business climate surrounding intellectual property, the Harvard Business School published a fairly popular book entitled *Rembrandts in the Attic: Unlocking the Hidden Value of Patents.* The authors praised the entrepreneurial spirit of Eugene

Emmerich, CEO of Cadtrack, a company formed to produce computer workstations. Cadtrack was headed toward bankruptcy in 1983 when the head of licensing at IBM called to discuss the possible licensing of a patent for moving a cursor on a screen. In 1985, Emmerich went to the board of his company and suggested that the company get out of the production business. The company then laid off all of its employees and concentrated on collecting revenues from its patent. By 1997, when the patent finally expired, he had signed deals worth about $50 million with 400 companies.[57]

Emmerich was proud that only one company refused to take Cadtrack's license—Commodore. He boasted, "So we took them to court and got a permanent injunction barring sales of their computers in the U.S. When that happened, their creditors called in their loans and they went bankrupt. That little patent of ours put Commodore out of business."[58]

So here was a failing company that won an excessively broad patent for something that many firms had been doing before it won the patent. It made no contribution to the advancement of technology. Instead, it merely collected royalties, much to the irritation of those who followed the industry at the time.[59]

"Submarine patents" perhaps represent the most blatant abuse of the patent system. In effect, an "inventor" with no intention of actually creating anything useful submits a patent application with vague ideas—the broader, the better. Then the "inventor" lies in wait until a person or company develops an idea that bears some similarity to the initial patent. At that time, the patent holder threatens to sue. The actual inventor may pay an exorbitant amount just to avoid the cost of litigation and/or the risk of being found guilty of patent infringement.

George P. Selden, a patent lawyer, developed what might be the most famous example of a submarine patent. He kept a patent application for the automobile alive from 1879 to 1895 by continually revising it. Selden never intended to build a working vehicle, but only to collect royalties. The key claim of his patent was the use of light gasoline to fuel an internal combustion engine to power an automobile.[60] Eventually, the Electric Vehicle Company gained control of the patent, intending to use it to discourage the production of gasoline cars or, if it failed in that objective, to collect sizeable royalties. Henry Ford refused to recognize the patent and finally won a long, hard-fought court case.

The most important modern practitioner of the art of the submarine patent was probably the late Jerome Lemelson, of whom somebody wrote that he invented patents rather than patenting inventions. Lemelson claimed more than 500 patents on technologies used in industrial robots and automated warehouses, as well as fax machines, VCRs, bar-code scanners, camcorders, and the Sony Walkman.

Despite his wide-ranging patents, I do not know of any inventions that he developed other than the few children's toys with which he began his career.

After he saw his ideas for toys stolen, he vowed to take up the cause of independent inventors.[61]

Rather than actually developing working models of technologies, Lemelson merely filed patents to position himself to lay claim to advances that others developed independently of his ideas. Once somebody developed a product that had some relation to one of his patents, Lemelson would confront the would-be violator of his intellectual property, demanding compensation. Either to avoid the legal costs or because the courts might even hold for the plaintiff, Lemelson, as the holder of the submarine patent, would collect royalties or some other form of payment.

In email correspondence, I once asked Lemelson's son, Eric, if any of his father's inventions actually furthered technical progress. He responded that his father approached numerous corporations with his ideas. The one specific example he offered suggested that, at least in that one case, a corporation possibly may have made some use of one of his father's ideas. Lemelson had patented the idea for electronic locks on guns. Although none of the manufacturers seemed interested at the time, only one week after his patent expired, a manufacturer announced that it was going to produce such a gun. Even in this instance, given the current concerns over gun violence, both accidental and intentional, the company may have forgotten about Lemelson and developed the idea on its own.

Lemelson called his group of inventions covering machine vision his "greatest accomplishment as an inventor."[62] Here's the abstract of U.S. Patent No. 5,351,078 as it was issued in 1994, entitled "Apparatus and Methods for Automated Observation of Objects":

> Apparatus and methods are disclosed for automatically inspecting two- or three-dimensional objects or subjects. A detector and the object are moved relative to each other. In one form, a detector, such as a camera or radiation receiver, moves around an object, which is supported to be rotatable such that the detector may receive electromagnetic energy signals from the object from a variety of angles. The energy may be directed as a beam at and reflected from the object, as for visible light, or passed through the object, as for x-ray radiation. Alternatively, the detector passively receives energy from the object, as in an infrared detector. The detector generates analog image signals resulting from the detected radiation, and an electronic computer processes and analyzes the analog signals and generates digital codes, which may be stored or employed to control a display.[63]

The idea is interesting, and perhaps even novel, but does not seem to further technology any more than the ideas about air travel found in a Jules Verne novel or a Leonardo da Vinci notebook. Certainly, many of the companies that Lemelson sued questioned whether his patents merited the protections granted to owners of intellectual property. Eric Lemelson reported that in one case, Mitsubishi put his father through 66 consecutive days of depositions. Given the stakes involved, such efforts are not surprising. They are, however, exceedingly wasteful.

In the end, Lemelson's strategy proved highly lucrative. In 1992 alone, his lawyer, Gerald Hosier, claims to have closed deals for Lemelson worth $400-450 million.[64]

The wasteful litigation associated with Lemelson's intellectual property claims continues today, even after his death. A partnership representing Lemelson's interests continues to sue to win still more royalties. Now, however, others are suing Lemelson's interests claiming violations of their own intellectual property rights. Cyprus Amax Mineral Co., which is now in the process of merging with ASARCO to form Asarco Cyprus, claims ownership of the rights to all of the patents for which Lemelson applied between 1954 and 1957. According to the company, Lemelson signed an agreement that assigned to United States Metals Refining Co. all patent rights that he developed while he was an employee of the firm between 1953 and 1958. Because that firm became a part of Cyprus Amax Mineral Co., it now claims to own Lemelson's patent rights for inventions patented around that time.[65]

At least Lemelson did come up with concepts on his own. Other companies rummage around in the debris of failed businesses looking to buy up inexpensive patents that have gone unnoticed. One such enterprise, TechSearch, is demanding licensing fees for a number of its 25 patents, involving such areas as online education, computer-chip "emulation," and the ability of computers to recognize when peripherals such as printers are plugged into them—all of which have long been in common use. So far, one patent has brought in at least $2 million in fees from companies that believed that paying TechSearch would be cheaper than defending a case in court.[66]

TechSearch purchased a patent on chip emulation from since-bankrupt Texas chip developer International Meta Systems in January 1998. It paid $50,000 plus 10 percent of future recoveries. TechSearch then took the patent and tried to work out a licensing deal with Intel. Intel refused to pay, so TechSearch sued Intel, requesting damages of roughly $500 million. After losing its case, the judge ordered TechSearch to pay Intel $132,500 in legal fees. TechSearch has appealed. TechSearch is also suing Intel for libel and slander because an Intel spokesman remarked to the *Wall Street Journal* that TechSearch "exists solely for the purpose of purchasing patents and extorting funds from another company."[67] The case became even more Byzantine when Intel used a shell company in the Cayman Islands to attempt to void the sale of the patent to TechSearch.[68]

Most companies would rather avoid the expense of a court trial. The American Intellectual Property Law Association estimated that as of 1998, the average cost of a trial ran about $1.5 million per side. Since TechSearch demands less than $100,000 for a license, settling seems prudent.[69]

Of course, this practice does nothing to promote research. In fact, it impedes research by requiring companies either to fork over money or become entangled in lengthy litigation.

A Russian firm, Intellect, went one step further than Lemelson. It recently won a Russian patent for bottles: Not just for a particular design of bottles, but bottles in general—any kind of bottle. The company by its own admission has never made a bottle. It merely submitted a sketch of a bottle to the Russian patent office. Now that it owns the exclusive right to make bottles in Russia, it has begun demanding royalties from Russian brewers and even Coca-Cola for the use of its intellectual property in Russia.[70] In the long run, the owners of many of the more ridiculous patents will be unable to enforce them, but even so huge sums of money will be wasted in the effort to sweep away such abuses.

Certainly, Lemelson milked the system and grew enormously wealthy in the process. The moral of Lemelson's career, however, is more about the absurdity of the patent system than the actions of a particular individual.

In one respect, what Lemelson did makes considerable sense. Perhaps the only course for an independent inventor is to follow his example: file patents and sue. Within the present system of intellectual property, independent inventors who either try to develop technologies on their own or attempt to interest a corporation in a collaborative effort are at a substantial disadvantage. The giant corporations can call upon powerful legal firms that employ well-connected political operatives as well as first-class lawyers. I will discuss one of these firms—Aiken, Gump—in a later section.

Patenting Business Practices

Recently, the Patent and Trademark Office has moved onto another slippery slope that will make the problems associated with the Lemelson patents pale by comparison. It has begun to grant intellectual property rights for business models; that is, for a way of doing business. Yes, a person can actually patent an idea for a way of doing business. For example, Jay Walker, the billionaire founder of Priceline.com, aims to become the Jerome H. Lemelson of the Digital Age. He has built a virtual idea factory in Connecticut, which he calls Walker Digital.

Walker Digital does not just develop ideas; it also churns out patents. As Seth Shulman, a journalist who has closely followed the company, observed, "Fully a third of Walker's 60 employees staff its legal department: electronic scribes who spin out an average of two highly conceptual patent applications every week."[71]

Walker Digital has already won 50 patents for business practices, and another 300 are pending, according to the company's Web site.[72] For example, the patent for Priceline.com covers the process of taking online bids, in which companies respond to customers' declaration of a willingness to pay for their service or a product. Walker Digital also has a patent for a new method of selling magazine subscriptions[73] and another for a concept involving the online buying and selling of expert opinions.[74] A company that stumbles upon these same ideas will have to compensate Walker for using his intellectual property rights.

Granting patents is supposed to stimulate innovation. How innovative was Priceline's original patent? On October 20, 1997, William J. Gurley published a column, which was later published nationally in *Fortune,* outlining a service remarkably similar to Priceline seven months before the company was launched and with no knowledge whatsoever of Priceline.com's business.[75]

More seriously, a San Francisco company called Marketel filed a suit against Walker and Priceline, alleging that Walker stole its trade secrets. Marketel began BookIt, an airline ticket reservation business in 1991, six years before Priceline began. The Marketel model was very similar to that of Priceline, except it used faxes instead of the Internet. Customers sent in requests for airline tickets by fax, guaranteeing their bid with a credit card. If an airline was willing to sell a ticket for the price that the customer bid, the deal went through. Marketel says that in 1987 William Perell, the company's president, hired a friend, Andre Jaeckle, to find investors for the BookIt business. According to Jaeckle, he introduced Walker to Perell in the fall of 1988. Over the next 18 to 24 months, according to Marketel, Walker, Perell, and Jaeckle talked about the BookIt business plan, and Perell revealed details of its workings.[76] However, a district court ruled against Marketel on the narrow ground that Walker's nondisclosure agreement was only for one year, although the decision is open to appeal.

Rather than stimulating innovation, these patents merely protect business from competition. For example, Amazon.com is currently suing the Barnes and Noble online book-selling affiliate. Amazon claims that its methods of selling over the Internet are proprietary. Wal-Mart, in turn, sued Amazon, claiming that Amazon pirated some of its technology. In the settlement, Amazon had to reassign eight former Wal-Mart employees accused of revealing proprietary information.

Obviously, the Barnes and Noble operation is imitating Amazon.com, but businesses have always done so. If a Mexican restaurant becomes popular in a town, other Mexican restaurants are sure to follow. The patenting of business practices makes that strategy difficult to pursue. The situation would be comparable to a Mexican restaurant patenting tortillas or tamales to prevent competition from other Mexican restaurants. The rivals might complain that people have been making these foods for as long as anybody can remember, but first they must prove their case in court. They must convince the authorities that the recipes have been common practice long before the granting of the patent. They might prevail in the courts, but the time and expense could be so great that they might not think the effort worthwhile.

While nobody to date has succeeded in patenting tamales, one Mexican restaurant in Houston has already won a suit against another restaurant for copying its decor (Fisher 1999). The combination of legal costs and fees for intellectual property rights is likely to prove deadly for commerce if present trends continue.

Consider the case of Ed Pool, who won a patent for using the Internet for computer-to-computer international trade. He intends to collect 0.3 percent of each transaction. If he succeeds in enforcing his patent, he will earn billions and billions of dollars per year.[77] Mr. Pool did nothing to help facilitate such transactions. He merely filed a patent application that will permit him to sue people for what they were already doing, setting off another round of wasteful litigation.

A recent piece in *Business Week* described "Intellectual property as the Web's war zone," noting: "Since 1995, federal lawsuits over patents, copyrights, and other intellectual property have risen 10 times faster than other cases brought under federal law, topping 8,200 cases last year."[78] A business-friendly columnist in the *Wall Street Journal* referred to the intellectual property claims of business methods, such as Amazon's patent for 1-click checkout and Jay Walker's Priceline.com, as "borderline idiotic."[79] Another article in the same paper expressed the growing alarm over the abuse of the patent system by holders of intellectual property:

> Suddenly, competition in the new economy is shifting from the Internet into the courtroom. The biggest names in e-commerce are suing each other over patents, and roiling the Web in the process. Until now business titans of cyberspace have railed against any form of government meddling. But the Net is growing up, and companies like Amazon want patent examiners and judges to intervene. The battles brewing over these patents threaten to slam the brakes on the Web's breakneck pace of innovation. Clearly Internet companies have a right to protect their intellectual property. But if a few big companies control key methods of doing business online, future start ups could find themselves

> spending more time navigating the legal thicket and less time creating the next new thing. . . . Instead of circuits or gears, these patents cover methods and processes. Critics say they amount to little more than Web retreads of old business ideas.[80]

Tim O'Reilly, president of O'Reilly & Associates, a leading computer publisher, goes even further. He compares some of the current patents for business practices to a young medical student winning a patent on the mere notion of a cure for cancer, without actually developing any medicine to accomplish that objective.[81]

To make matters worse, the U.S Patent and Trademark Office is woefully unequipped to handle the flood of patent applications for business methods. The agency has just 38 examiners considering such applications, the volume of which doubled in 1999 to nearly 2,700, or an average of more than 71 applications per examiner per year. This caseload, which gives little more than 3 days on average for an examiner to rule on a patent, will get even heavier. The total for the year 2000 could reach 6,000, 65 percent of which will likely be approved.[82]

The Law as Intellectual Property

Intellectual property rights obviously depend upon the law, but until recently, one company virtually claimed the law itself as an intellectual property right. West Publishing Co., of Egan, Minnesota, had a contract with the government to print the federal case law. Like most books, these law books have page numbers to allow people to locate specific passages more easily.

Today, lawyers increasingly rely upon computer-assisted legal research to access prior legal decisions throughout the nation. Two companies dominate online access to legal information, West Publishing's Westlaw and Reed Elsevier's Lexis-Nexis. This market is extraordinarily lucrative. A single large law firm can spend millions of dollars in fees annually for online research.

West built its own database in the 1970s and 1980s by sending its law books to Southeast Asia and the Caribbean, where low-paid clerks laboriously key-punched the information. Lexis and West litigated for years about whether West could copyright its page-numbering system, which many courts have long required attorneys to use in legal papers. Finally, in 1988, Lexis agreed to pay a license fee to West for its page-numbering system.[83]

More recently, two smaller companies, Matthew Bender and HyperLaw, wanted to compile case law on CD-ROMs to offer a relatively inexpensive alternative to the online services. West challenged these competitors, charging that they had not merely reproduced legal cases. Again, West based its claims on its system of page numbering. The company maintained that although the legal decisions might be in the public domain, its numbering of the pages added value to the material. Consequently, the corporation was justified in copyrighting its page numbering.

According to West's position, vendors of CD-ROMs could still publish the cases, but they could not indicate which text fell on which page of the printed version. In 1995, West attempted to get Congress to give legislative authority for its copyright under the so-called "Paperwork Reduction Act." This attempt failed. West also sued Matthew Bender and HyperLaw for copyright infringement. It lost that suit as well in a federal appeals court in 1997. Reed Elsevier eventually swallowed up Matthew Bender, but HyperLaw remained vulnerable to West. Two years later, the Supreme Court finally refused to hear West's appeal.

Before celebrating the ultimate legal outcome, keep in mind that such legal challenges are expensive for a small company. Also, consider that the decision of the appeals court was a split decision: one of the three judges agreed with West. The outcome of the legislative effort failed only because of an intense lobbying effort. Finally, the favorable outcome was probably not unrelated to the fact that lawyers as a group had a vested interest in this case. Many are consumers of the expensive services of West Publishing or its competitor, Lexis-Nexis, and would most likely appreciate the existence of competitors that could drive down the prices of computer-aided legal research.

The fact that West could even consider claiming the page breaks in a text as intellectual property suggests how absurd the whole system has become. During the early stages of this dispute in September 1994, Bob Oakley, the Washington, D.C. representative of the American Association of Law Libraries, wrote to U.S. Attorney General Janet Reno: "[I]t is a fundamental part of our belief that no one should own the law, either outright or in practical effect. . . . Regrettably, the assertion of ownership of some parts of the published case law together with the requirements of courts and others to cite certain privately published versions of the case law have, in practical effect, given one publisher substantial control over the legal information market."[84]

From another perspective, West Publishing may have had a point. Since the law seems to support so much nonsense regarding intellectual property, treating the law itself as intellectual property might be fitting. The added expense and inconvenience for lawyers who need to access the law as private property could act as a useful reminder to lawyers and judges about the need for caution in pushing intellectual property rights in directions that nobody, until recently, had even imagined.

In other cases, one would have to search harder to find a justification. California and 47 other states have building codes that are copyrighted by one of three nonprofit organizations. The federal government requires U.S. physicians to use a medical billing code that's owned by the American Medical Association. The National Fire Protection Association's copyrighted 900-page electrical code is in force, in one form or another, in all 50 states, plus Puerto Rico and Guam. In all of these cases, people are expected to follow public laws that are also private intellectual property. A printed copy of the California Building Code costs $738.[85] Legal challenges are underway, but they have not succeeded so far.

Hypocrisy and Intellectual Property

Corporate holders of intellectual property are quick to call for government action to prevent others from violating their sacred intellectual property rights. Their prevailing rhetoric proclaims the absolute sanctity of intellectual property. Yet, judging by a growing number of court cases, their respect for intellectual property rights is not without limits. Apparently, while expressing righteous indignation when their own intellectual property rights are violated, these same companies frequently resort to illegal techniques to steal intellectual property from others.

For example, competitors have time and time again brought Microsoft to court charging violations of intellectual property or antitrust abuse. Microsoft has either lost or settled out of court a number of these suits, including Caldera, Stac Electronics, and Bristol. In a more recent dispute, Kodak had been working with Microsoft for a year on a new photo-transfer standard that allowed Windows to recognize cameras plugged in to a computer. A front-page article in the *Wall Street Journal* reported that when Kodak technicians received a copy of Microsoft's new operating system, they were shocked by what they saw:

> When Kodak cameras were plugged into a PC loaded with Kodak software, it was Microsoft's own photo software that popped up—not Kodak's. Camera customers would have to go through a cumbersome process to get Kodak's software to pop up every time, and most would probably just use Microsoft's. More troubling, the Kodak team found that the new program steered orders for picture prints to companies that would have to pay to be listed in Windows, and that these companies also would be asked to pay Microsoft a fee on every photo sent through Windows.
>
> Kodak's story offers a snapshot of a now-familiar tale in the software business.[86]

Under intense unfavorable publicity and under the shadow of a continuing antitrust suit, Microsoft eventually relented and agreed to accommodate Kodak. Taken together, the allegations and complaints suggest a pattern in which Microsoft proposes or undertakes to negotiate some sort of partnership, then learns about the new technology through the "partnership," and finally attempts to use that information to create a product that competes with and then threatens to destroy its erstwhile partner. Microsoft might respond that its deep pockets rather than its misbehaviors motivate many of these allegations, but the pattern of claims does raise red flags.

Even more blatantly, Microsoft copied the "look and feel" of Windows from Apple's Macintosh operating system, at a time when intellectual property rights were admittedly weaker. Apple complained, but earlier it had lifted much of the Macintosh system from Xerox's Palo Alto Research Center.

While the corporate beneficiaries of intellectual property rights pretend that they base their justification of their "ownership" of intellectual property on the incentives that it creates for the future development of science and technology, their private communications suggests a less flattering motivation. For example, on May 16, 1991, Bill Gates wrote a confidential memo to senior Microsoft executives that revealed the real nature of the modern patent process. He warned:

> If people had understood how patents would be granted when most of today's ideas were invented and had taken out patents, the industry would be at a complete standstill today. I feel certain that some large company will patent some obvious thing related to interface, object orientation, algorithm, application extension, or other crucial technique. If we assume this company has no need of any of our patents, then they have a 17-year right to take as much of our profits as they want. The solution to this is patent exchanges with large companies and patenting as much as we can. Amazingly we haven't done any patent exchanges that I am aware of. Amazingly we haven't found a way to use our licensing position to avoid having our own customers cause patent problems for us. . . .
>
> A future start-up with no patents of its own will be forced to pay whatever price the giants choose to impose. That price might be high: Established companies have an interest in excluding future competitors.[87]

In other words, although the patent system is defective for furthering technical progress, it will serve the corporate interests by protecting established companies from competition if only they know how to play the patent system.

On the public stage, Microsoft and other major corporations still proclaim the sanctity of intellectual property as a way of securing the maximum rate of technical progress, which will supposedly benefit each and every one of us. Ironically, in its internal operations, Microsoft not infrequently uses free, open-source software rather than rely on its own products for challenging operations.[88]

Like Microsoft, Monsanto has a well-deserved reputation of rigorously enforcing its intellectual property rights. Like Microsoft, it is also less enthusiastic about the property rights of its competitors. For example:

> In 1994, the company Agracetus was awarded a European patent which covered all genetically engineered soybeans. Rival companies, including Monsanto, were outraged and immediately challenged the patent, saying that it would result in just one company having an effective monopoly over all transgenic soybeans. Monsanto argued that "the alleged invention lacks an inventive step and was not . . . novel." In the end the solution for Monsanto was to buy Agracetus, together with the patent, and drop the complaint. As well as the patent on Soya, Monsanto now holds a patent in both Europe and the US on all genetically engineered cotton.[89]

You might not be surprised that Monsanto no longer considers such a broad patent to be outrageous.

The Confiscation of German Intellectual Property

In the late nineteenth and early twentieth centuries, Germany had the most advanced educational system in the world. Germany had no uniform system of patents, although Germans took out patents in Britain and elsewhere. Although the patent system is supposed to be the keystone of economic progress, the German educational system produced magnificent technological achievements.

The German industrial lead was probably greatest in the production of chemicals. During World War I, Britain, France, and the United States confiscated all of Germany's patents. Even then, companies in Britain and the United States often were unable to develop the processes described in the patents without purchasing trade secrets from the Germans.

The Americans took nine years to discover that among the patents seized during the war were processes to manufacture synthetic methanol, silk, cotton, and wood. By that time, the Germans had developed new procedures.[90] Even so, on the basis of their expropriated intellectual property, Britain and the United States were eventually able to challenge the German monopoly in the production of chemicals.[91]

History repeated itself after World War II, when the Allies expropriated Germany's intellectual property once again. Over and above the outright confiscation of German patents and internal documents, the United States captured a good number of German civilians with skills in research and technology, placing them in the custody of American corporations.[92] Crippling the German economy might have seemed to be just retribution, but the taxpayers of the United States then had to foot the bill for supporting the German recovery. So, in effect, the government paid for the accumulation of intellectual property rights by private corporations.

Further Strangeness Concerning Intellectual Property

Much of the effort in collecting patents has little to do with the protection of vital research that is essential to technological progress. Frequently, as the Gates memo illustrates, companies intend to accumulate a strategic set of patents merely to block potential competitors from advancing.

On a more humorous level, Napster, the company that facilitated the sharing of music over the Internet, sent a cease-and-desist order to a punk-rock band, the Offspring, when it started selling T-shirts featuring the Napster logo.[93] In other words, Napster's own intellectual property was sacrosanct.

The U.S. government has been at the forefront of those advocating the protection of intellectual property, yet some indications suggest hypocrisy is at work here as well. At least some agencies of the government seem to be less wedded to the sanctity of intellectual property than the official rhetoric would suggest. For example, a German company, Enercon, had developed a new design for the generation of electricity from wind. When the company tried to market its invention in the United States, its American rival Kenetech announced that it had already patented a nearly identical system. Kenetech then sued to ban the sale of Enercon equipment in the United States, in an effort to keep the market to itself.

Although Enercon had developed the variable speed rotor in the 1980s, Kenetech's initial public stock offering implied that it was the prime innovator of variable speed rotors. The Patent and Trademark Office did not question this claim and awarded Kenetech a very broadly written patent. Nonetheless, the company filed for chapter 11 bankruptcy on May 29, 1996.[94]

In a rare public disclosure, an employee of the U.S. National Security Agency (NSA) agreed to appear anonymously in silhouette on German television in August of 1998 to reveal how he had stolen Enercon's secrets, using satellite information to tap the telephone and computer link lines that ran between Enercon's research laboratory near the North Sea and its production unit some 12 miles away. Detailed plans of Enercon's supposedly secret invention were then allegedly passed on to Kenetech.[95]

Enercon lost its initial encounter in court. That outcome does not necessarily invalidate the possibility of clandestine activity. The NSA is perhaps the most secret arm of the government of the United States. Proof of accusations, such as Enercon's, is virtually impossible. The clandestine operatives of the NSA would certainly be expected to do anything possible to keep any traces of their activity from the public eye. If you were on a jury, which side would you believe?

The Enercon allegations are far from unique. A report for the director general for research of the European Parliament compiled a number of similar instances in which U.S. government intelligence agencies gathered private information to support the needs of private corporations.[96] For example, the report noted that the NSA supposedly used its sophisticated eavesdropping capacity to monitor all the faxes and phone calls between the European consortium Airbus, the Saudi national airline, and the Saudi government. The agency found that Airbus agents were offering bribes to a Saudi official. It passed the information to U.S. officials to assist Boeing Corp. and McDonnell Douglas Corp., which eventually won the $6 billion competition.[97]

After some of these revelations came to light, James Woolsey, former head of the Central Intelligence Agency, responded dismissively: "Most European technology just isn't worth our stealing." Instead, he justified the actions of the NSA on the grounds that the Europeans were supposedly bribing other countries to purchase their technology.[98] Of course, in major bids most companies engage in bribery even though the law prohibits it.

This case suggests another dimension to the hypocrisy of the claims for intellectual property. Recall that one supposed purpose of the patent system is to make technical information available to the public in the form of patent applications in return for the exclusive right to profit from the information for a period of time. The NSA secured U.S. Patent No. US5937422 on August 10, 1999: "Automatically generating a topic description for text and searching and sorting text by topic using the same."[99]

Presumably, this patent covers the technology that allows the NSA to filter telephone conversations, faxes, and emails, searching for keywords that may indicate a subject that concerns the agency. So here is a top-secret agency publicly requesting protection of its intellectual property so that it can, among other things, violate the privacy, if not the intellectual property rights, of others.

Concluding Note

None of these problems could have occurred to the framers of the Constitution when they were drafting their provisions for intellectual property. Recall the earlier discussion of the sparse system for creating new technologies. In an environment in which relatively few inventions were coming on line and many fields had virtually no domestic inventions, they would have been unlikely to foresee the confusion and litigation that arise with conflicting claims and a threadbare system of patent examination. In the present situation, intellectual property laws have become destructive. As the claims to intellectual property expand, the resulting conflict can only become worse.

CHAPTER THREE

INTELLECTUAL PROPERTY VS. SCIENCE

Eating the Scientific Seed Corn

The intense efforts to protect the sanctity of intellectual property are ironic. Although the promoters of intellectual property justify their claims on the basis of the wonders of new technology, the flowering of today's commercial successes depends on scientific discoveries from decades ago, when science still enjoyed considerable public funding. However, the private interests that now stake claims for intellectual property rarely recognize how much this earlier work, centered in universities or government agencies, enriched the public domain of science and technology.

In developing their technologies, most corporations have looked to research from the universities or the military, with the exception of a few special corporate research environments, such as Bell Labs or the IBM research centers. To make sure that I was not overlooking something in this regard, I recently asked participants in a business history mailing list for examples of revolutionary technologies developed by corporations without funding from government sources. I excluded Bell Labs and the IBM research centers. Usually such requests elicit a broad set of responses. In response to my question, the only technology mentioned was the catalytic cracking of petroleum. So, I feel confident in my claim that while corporations may have successfully commercialized previous research, they depended on science and major inventions from outside of the corporate sector.

In a major study of recent patent applications, Francis Narin and his colleagues attempted to track down the funding source of the scientific research that the patent applicant cited on the first page of the application. They found that 73 percent of the main science papers cited by American industrial patents in two recent years were based on domestic and foreign research financed by government or nonprofit agencies. Even IBM—famous for its research prowess and numerous patents—was found to cite its own work only 21 percent of the time.[1]

An unpublished 1968 National Science Foundation study entitled "Technology in Retrospect and Critical Events in Science" traced the key scientific events that led to five major innovations: magnetic ferrites, videotape recorders, the oral contraceptive pill, electron microscopes, and matrix isolation. In all five cases, nonmission research—research motivated by the search for knowledge and scientific understanding without special regard for application—played a key role, and the time lag was significant, between 20 and 30 years. The study also found that 76 percent of the nonmission research was performed at universities and colleges.[2]

Similarly, James Adams, an economist at the University of Florida, measured the stock of knowledge available in a field by counting publications in the field over a long period, usually beginning before 1930. Then he created what he called industry "knowledge stocks" by multiplying these counts by the number of scientists employed in each field in each of 18 industries that he studied. Next, he analyzed productivity growth in each of these industries over a 28-year period, both from the industry's own knowledge stocks and from the knowledge stocks that flowed from other industries. He found that both types of knowledge stocks were major contributors to the growth of productivity.[3]

Not surprisingly, Adams found that the stocks of knowledge took a long time before they contributed to industrial growth. Knowledge stocks within an industry have an effect after about 20 years. Knowledge stocks from other industries take about 30 years before they contribute to productivity growth in another industry, in line with the estimates of the lags in the National Science Foundation study.[4]

These academic studies confirm my contention that, although corporations may succeed at commercializing basic research, they are ill suited for developing the basic science on which economic progress depends. Corporations generally want research for a specific purpose. In contrast, one provocative student of the subject proclaimed, "Basic research often provides answers to unposed questions."[5]

The current attempt to make science more practical has an interesting precedent—interesting especially because the recent attempt to make science more entrepreneurial is supposed to reflect growing confidence in the market. In the socialist economies of Eastern Europe, political leaders also attempted to make scientists' work more practical. In the German Democratic Republic, for example, the national science academy earned half its income from industrial research contracts. One scholar attributed the slowdown in productivity growth in Eastern Europe to this attempt to control basic science, concluding: "In this general context, the fate of research in the socialist academies of sciences might sound a warning note" to those who applaud the recent entrepreneurial turn of academic science in the United States.[6]

The Dying Tradition of Scientific Openness

Over and above the direct costs of intellectual property rights, the ascendancy of intellectual property presents a grave threat to the vigor of the scientific process. Historically, science has been a collaborative process in which multitudes of people individually contribute a part to a cumulative process. This scientific ethos is antithetical to the current system of intellectual property, which depends on a single agent claiming credit for the entire process. As Robert Merges and Richard Nelson have written:

> [T]he advance of public science is continuously illuminating new technological opportunities. Particular inventive efforts . . . often can yield very large advances in technological capabilities over what had been prior best practice. Yet the contribution, in economist's jargon the value added, of the "first to bring to practice" may be quite small, in that the direction to go was "obvious," and if he or she hadn't achieved success, someone else would have very soon. Nor in many of these cases were the expenses and risks involved in the winning efforts so great, or anticipated to be so, that only the expectation of a giant reward could have induced them. And the granting of a broad patent to the first to bring to practice cannot be justified as likely to lead to more effective and orderly development of the prospect, if history be our guide.[7]

I will discuss the discovery of the BRCA1 gene, which is related to the onset of breast cancer, in the next chapter. This event illustrates how this competition to be the first to get credit for a discovery destroys the sort of cooperation upon which science has depended historically. In a more dramatic example, on February 14, 1876, Alexander Graham Bell's father-in-law submitted Bell's patent for "Improvements in Telegraphy" just hours before Elisha Gray applied for a patent caveat, outlining his idea. Gray was a cofounder of Western Electric, which later became a part of the Bell System. Had Gray arrived a few hours earlier, the course of business and technology might have been quite different.

Aside from such accidents, the rush to be the first to claim intellectual property rights has a chilling effect on scientific openness. Prior to the time that the quest for intellectual property rights came to dominate science, ideas cross-fertilized each other. This process created an unparalleled explosion of scientific progress. Paul David, an economic historian with a specialty in science and technology, has shown how this success of Western science owed a great deal to a historical accident that created a tradition of scientific openness. In David's words,

> kings and nobles were immediately concerned with the ornamental benefits to be derived by their sponsorship of philosophers and savants of great renown. . . . [The] formation of a distinctive research culture of open science was first made possible and, indeed was positively encouraged by the system of aristocratic patronage.[8]

During the Renaissance, science progressed to the point where the aristocrats could no longer evaluate the scientific worth of their beneficiaries. As a result, they sponsored scientific societies that could reveal their relative merits to their benefactors. In this way, "the patronage system in post-Renaissance Europe induced the emergence and promoted the institutionalization of new reputation-building proceedings; these entailed the revelation of scientific knowledge and expertise among extended reference groups that included 'peer-experts.'"[9] This openness was the key to the explosive success of Western science.

This tradition continued for centuries, with great scientists and mathematicians vying with one another to attain glory for their scientific achievements. In this milieu, fame and recognition, especially among their peers, constituted the major payment for scientific discoveries.[10] Although recognition can occasionally lead to financial rewards, such as a Nobel Prize, more often it takes more mundane forms, such as associating the name of the discoverer with the discovery or even more commonly in the form of a publication.[11]

Given the emphasis on recognition, scientists attempted to communicate their discoveries whenever possible, at least once they had established the priority of their work.[12] So, rather than dashing to the patent office to claim their discoveries as private property, traditional scientists would rush to publicize their work in scientific journals.

Robert King Merton, a sociologist who pursued this subject with great vigor, concluded, "The substantive findings of science are a product of social collaboration and are assigned to the community. They constitute a common heritage in which the equity of the producer is severely limited."[13] In this vein, even the greatest scientists acknowledged that they were a part of a larger process rather than the sole discoverer of a new scientific principle. Sir Isaac Newton, who himself stacked a Royal Society committee to ensure that it would credit him rather than Leibnitz with inventing calculus, displayed this spirit of generosity in famously writing to Robert Hooke (February 5, 1676), "If I have seen further it is by standing on the shoulders of Giants." In fact, Merton wrote an entire book based on this letter.[14]

Spillovers

Many of the greatest technological achievements occurred because of scientific innovations unexpectedly jumping from one industry to another. Of course, to the extent that traditional science has been relying on military largess, especially in postwar decades, secrecy has become increasingly common.

Even in the military, where secrecy is supposed to be of the utmost importance, the possibly inadvertent sharing of ideas can be extremely productive. Consider the case of radar, which was at the cutting edge of military technology at the time of its development. Melvin Krantzberg, a noted historian of technology, observed: "Radar . . . is a very complicated system requiring specialized materials, power sources, and intricate devices to send out waves of the proper frequency, detect them when they bounce off an object, and then interpret them and place the results on a screen."[15] Because of the multitude of crucial technological requirements, many people have claimed to be the inventor of radar.

In 1970 the Pentagon produced a study that found that "the U.S. lead in microwave electronics and in computer technology was uniformly and greatly raised after the decision in 1946 to release the results of wartime research in these fields."[16] The same study said nuclear reactor and transistor technology development also benefited from an open research policy. In fact, another scholar concluded that semiconductors were "the major offspring of radar."[17]

While spillovers from military-oriented science were common in the past, they are less so today.[18] Yes, computers, the Internet, and many other elements of high technology emerged out of military projects. Even so, most modern military products are just unsuited to the civilian sector. This lack of civilian applications is becoming more extreme each year.

For example, the early jet engines that the military developed had obvious civilian benefits. Once military planes entered the world of supersonic speeds, however, "they began to assume performance and cost characteristics that were inappropriate for the cost-conscious world of commercial travel." Similarly, communications satellites initially had great civilian spillovers, but the special requirements of contemporary military satellites are inappropriate for commercial communications.[19]

In addition, military projects that might seem to have some civilian application become distorted because of the specific needs of the military. For example, military projects often require electrical components that are "hardened" so that they can continue to function during a nuclear attack. Learning to "harden" electronic components does little to help society.

Obviously, the sort of public science I have in mind would not be militarily oriented. Instead, a community of scientists who have demonstrated their scientific talent and exhibited an intense commitment to a life of science would join together to try to unravel the mysteries of the world. These scientists would be absolutely open about their work with one another. In short, such a system would approximate the best aspects of academic science during its high point following World War II.

This system of public science has worked remarkably well over the past centuries, but today it is in jeopardy. Before I explain this danger, I will turn to a discussion of the development of the transistor, a centerpiece of the so-called New Economy.

Science—Defining Boundaries of Ideas

Assume for the moment that you think that intellectual property rights provide effective incentives to develop new technologies. How could you possibly administer such a system in a reasonable manner? If property rights are to be an effective organizing principle for organizing science and technology, they must be clearly defined. For example, my house has a door. People realize that they do not have the legal right to open the door and enter my house without my permission. In this case, my property right leaves relatively little room for ambiguity.

Even here, some ambiguity does exist. Other people can create noise or smells that invade my house. The fine print on my deed may even indicate that other people own the rights to any minerals that may lie beneath my house. Nonetheless, my house certainly has clear-cut boundaries compared with ideas.

Property rights in ideas are chock-full of ambiguity. After all, ideas do not emerge in a vacuum. People take inspiration from any number of sources. More often than not, they are not even fully aware of their sources of inspiration.

Nobody could possibly disentangle the complex pathways by which ideas develop. James Burke's television series and his monthly column in the *Scientific American,* both aptly named "Connections," illustrate the strange and unexpected sources of scientific progress. Even if somebody could somehow identify the various influences that contributed to an idea, nobody could determine the precise amount that each contributed to the final idea. Alfred Kahn, an economist mostly renowned for his role in promoting deregulation, wrote an excellent article on the patent system back in 1940, in which he described the impossibility of claiming the rights to an invention:

> Strictly speaking, no individual makes an invention, in the usual connotation of the term. For the object which, for linguistic convenience, we call an automobile, a telephone, as if it were an entity, is, as a matter of fact, the aggregate of an almost infinite number of individual units of invention, each of them the contribution of a separate person. It is little short of absurdity to call any one of the interrelated units the invention, and its "creator" the inventor.[20]

Just a few years later, Michael Polanvyi took Kahn's idea one step further:

> I believe the law is essentially deficient, because it aims at a purpose which cannot be rationally achieved. It tries to parcel up a stream of creative thought into a series of distinct claims, each of which is to constitute the basis of a separately owned monopoly. But the growth of human knowledge cannot be divided up into such sharply circumscribed phases. Ideas usually develop gradually by shades of emphasis, even when, from time to time, sparks of discovery flare up and suddenly reveal a new understanding, it usually appears on closer scrutiny that the new idea had been at least partially foreshadowed in previous speculation, moreover, discovery and invention do not progress only along one sequence of thought, which perhaps could somehow be divided up into consecutive segments.[21]

When intellectual property rights were less strong, businesses, as well as scientists, collaborated. For example Eric von Hippel, an economist at the Massachusetts Institute of Technology (MIT), has shown how producers of capital goods refine their products on the basis of the experiences and needs that their customers communicate to them.[22]

In a sense, science represents the very sort of spontaneous development that Friedrich A. Hayek, cited earlier as an opponent of the patent system, attributed to the market.[23] Of course, the spontaneous nature of scientific progress makes far more sense than spontaneous evolution of markets does. People have a fair idea of what they want the economy to do, especially in the near future, while science develops along a path full of surprises. People contribute to the scientific dialogue from unexpected disciplines, leading to unexpected results with unexpected consequences. Serendipitous discoveries from seemingly frivolous interests are the order of the day.

Once ideas become property, disputes over ownership become inevitable. Scientists and researchers frequently arrive at the same idea as other people without realizing that they are somehow violating their property rights. Of course, they could consult the Patent and Trademark Office at every step along the way, but to do so would impose a serious burden on their work. They would spend more time trying to avoid infringing the intellectual property of other people than creating anything. Should they, however, accidentally trespass on the intellectual property of another party, they will probably find themselves embroiled in lengthy litigation.

Even the most careful researcher will be unable to uncover all the potential pitfalls of intellectual property claims. Patents often include vague applications, so much so that the same invention may violate more than one earlier patent—even if the invention is novel. In the United States, the indeterminate nature of intellectual property rights is compounded by the fact that the Patent and Trademark Office has too few people with too little expertise to be able to clearly delineate who has a legitimate claim to intellectual property rights.

To make matters worse, the Patent and Trademark Office is extending the scope of patent rights into emerging areas that neither its examiners nor anybody else fully understands. As a result, a good number of people have become multimillionaires by exploiting the ambiguities associated with the paternity of ideas.

True, egregious abuses of the patent system can sometimes be overturned, but not before much damage is done. In the process, enormous quantities of time, effort, and resources dissipate in all sorts of wasteful struggles over intellectual property rights. To make things worse, small, innovative companies, and even more so individuals, will not stand a chance in a struggle with a major corporation over intellectual property rights.

The example of the radio, which represented cutting-edge technology in the early twentieth century, provides an instructive lesson about how destructive intellectual property rights can be.

The Transition to the Transistor

Earlier I mentioned the importance of Bell Labs and the IBM research centers. These facilities deserve some comment because they were so unlike the typical corporate research laboratory. In fact, they had more in common with the environment of a major university. Because their parent companies had been relatively immune from competitive pressures, they had accumulated immense profits. Perhaps most important, because the Bell System and IBM were so extensive, they could potentially reap enormous windfalls from their research.

For example, Bell Labs invented the transistor. The Bell System could profit from the savings of the transistor in perhaps thousands of locations across the nation. In contrast, a smaller company would have a limited demand for such an invention.

Because of the accumulated wealth of these companies and the potential internal demand for new technologies, researchers in these laboratories were relatively independent to pursue their own interests in basic research. Unlike most corporate researchers, scientists in these laboratories did not have to direct their research to a narrow subject with a quick payoff. These labs did not impose the sort of direct control common to most corporate research centers. The freedom enjoyed at these laboratories, of course, was not absolute. Management wanted to have some idea about the potential payoff, but it did not expect a guarantee of success and it realized that some unforeseen benefits could result.

These labs provided generous budgets for their projects. By providing freedom to pursue interesting studies and the opportunity to work with first-rate researchers, Bell Labs and IBM could hire elite researchers who would not consider working for other corporations.[24]

The history of the transistor, perhaps the crowning glory of the many discoveries that flowed from Bell Labs, illustrates the nature of this research environment. Bell Labs budgeted more than $1 million annually in the early years of this project. AT&T did not anticipate the transistor. It expected that this research would improve rectifiers or amplifiers. Nonetheless, the organization realized that AT&T had a vital interest in this area of research because of the potential benefits for communications technologies.[25]

The researchers who developed the transistor were not doing pure science. A 1931 paper had already laid out the basics of a quantum mechanical model of a solid semiconductor.[26] Their work was not merely applied, either. It was something in between. The history of the transistor also illustrates how important accidents can determine intellectual property rights. A team at Purdue University was within weeks of discovering the transistor.[27]

If AT&T had been free to use its intellectual property rights in the transistor the way contemporary firms can and do, modern technology would be far less advanced than it is today. Merges and Nelson have written:

> Because of an antitrust consent decree, AT&T was foreclosed from the commercial transistor business. . . . [As a result] AT&T had every incentive

> to encourage other companies to advance transistor technology, because of the value of better transistors to the phone system. AT&T quickly entered into a large number of license agreements at low royalty rates. Many companies ultimately contributed to the advance of transistor technology, because the pioneer patents were freely licensed instead of being used to block access.[28]

Because of government intervention, intellectual property rights did not limit the revolutionary potential of the transistor as they might have. Richard C. Levin, economics professor and later president of Yale University, speculated some time ago that the computer industry might not have developed if AT&T had not been forced to license the transistor to all comers.[29]

In reality, the Bell System was so big that AT&T might actually have done better by extracting less from fees and benefiting more from the resulting spread of new technology. A smaller operation would not have had that option. The key point is that the violation of AT&T's intellectual property rights was a seminal event in promoting modern technological advance.

Even so, the spread of the transistor was relatively slow. Scientists at Bell Labs invented the transistor in 1947, but for the next few decades the transistor had few applications. Only after a long time did this invention really bear fruit. As late as 1962, sales of vacuum tubes were still roughly double sales of transistors.[30]

So here is a technology based on a long history that began with scientific observations about the properties of semiconducting materials made at the beginning of the twentieth century. Researchers then followed with more theoretical work in the first decades of the century. Later, intensive military research for radar, as well as commercial research to develop the technology for the monopolistic telephone company, further nurtured the technology.

As was the case with the radio and the airplane, a violation of intellectual property rights substantially accelerated technical progress. The culmination of the work on the transistor was the computer chip, which now represents the centerpiece of the so-called New Economy. Although the corporations are commonly credited with the creation of this new economy, they really stood at the end of the pipeline, profiting from many decades of work that had nothing whatsoever to do with the competitive marketplace.

Unfortunately, this sort of misperception about the creative function of the corporations is common. The computer, which is the product of many decades of military research and funding, is typically regarded as the work of individualistic young men working in their garages. In other fields besides electronics, the corporations have prospered by privatizing this science while doing little to replenish the pool of basic science upon which their commercialization depends.

Society owes the great achievements taking place around us to the sort of science that developed in the post–World War II era, in which universities and the great research labs worked free from most commercial pressures.[31] With few exceptions, the corporate sector has not really developed any revolutionary technologies without substantial government funding or sponsorship.

Unfortunately, the political climate today has become contaminated with the illusion that the corporate model offers the ideal solution for all aspects of society. Basic university research has withered as the universities have become increasingly entrepreneurial.

Similarly, Bell Labs and the IBM research centers have become more focused on the immediate commercialization of technology rather than the basic science to which they contributed so enormously in the past. A year before the breakup of AT&T in 1981, Bell Labs had 25,000 employees and a $2 billion budget, of which roughly 10 percent was devoted to basic research. The business press worried that its performance would suffer in a more entrepreneurial environment—and with good reason.[32] While employment did not fall off, today IBM and Bell Labs both seem to measure their success in the quantity of patents they accumulate.

For example, a January 2001 IBM press release reads, "For the eighth consecutive year, IBM was awarded the most U.S. patents in 2000." Similarly, Bell Labs reports on its Web page, "Bell Labs is so productive it receives four patents a day on significant advances in communications." Patents, unfortunately, are a better indicator of market dominance than technical achievement. So, despite its pride in its patents, today, Bell Labs, which is now part of Lucent, is reduced to a mere shadow of its former self.

In this entrepreneurial world, who will fund the basic science that makes possible future inventions comparable to the transistor? Where are the future Bell Labs?

The Commercialization of the University

I do not want to give the impression that in the recent past universities in the United States stood wholly aloof from the commercial world. In reality, many universities have had close connections with the commercial world for a long time. Although religious denominations founded most of the early universities in the United States, they relied on business people, especially wealthy merchants, for a substantial portion of their funds. To promote contributions, the president of the school, typically a clergyman, would teach the courses in economics in such a way to make the institution seem attractive to the merchants upon whom the school would call for donations.[33]

Most applied scientific programs also grew up with a close connection to the corporate world. Programs in mining, engineering, and chemistry all had strong industrial roots. These historical connections between the university and business still resonate today. Think about the rise of Silicon Valley, home of the modern computer industry. The early mining industry played a crucial role in the evolution of this region. Because California lacked coal, early entrepreneurs and engineers dreamt of harnessing waterpower to generate electricity. The state's gold mining industry had considerable experience with hydraulic engineering. Stanford University had an exceptionally strong program in mining engineering. Herbert Hoover, perhaps the most famous mining engineer of all time, graduated from this program in 1891.[34]

The exploitation of this novel technology for generating electricity put Stanford, with its solid base in mining technology, in the forefront of early work in electronics. Shortly thereafter, the school established a new program in electrical engineering, which became an important element of the California electric-power industry within a decade of its opening. In addition, as was discussed in the second chapter, the San Francisco Bay Area developed a vibrant radio industry. In the first decades of the twentieth century, the radio industry worked closely with leading administrators and faculty members of Stanford.[35]

The maturation of this industry paralleled the career of Frederick Terman, who first worked in the local radio industry, then rose from a position as a professor of electrical engineering to dean of the engineering school, and still later became provost and vice-president of the university. Terman made a point of aiding his graduates in setting up businesses near the school. These businesses, such as Hewlett-Packard and Varian Associates, formed the core of the early Silicon Valley industrial district that later spawned the computer revolution.[36]

True, a massive infusion of government money was crucial to the creation of the computer industry, but much of the money went to Silicon Valley because of the infrastructure that had evolved out of the university's close connection with the mining industry.

The historian David Noble has detailed how the early growth and development of the Massachusetts Institute of Technology (MIT) was harnessed to serve the needs of science and industry.[37] These corporate ties generally persisted, even though some of the programs became more academic. Shortly after MIT set off along its path of corporate accommodation, Thorstein Veblen published *The Higher Learning in America: Memorandum on the Conduct of Universities by*

Business Men to protest the degree to which business influenced the academic world, based on his discouraging experience at the University of Chicago.[38]

People such as Andrew Carnegie and John D. Rockefeller were instrumental in advancing the business interest in higher education. Impressed with the ideas of Henry S. Pritchett, whom he heard during a White House luncheon in 1904, Carnegie began the Carnegie Foundation for the Advancement of Teaching, led by Pritchett. Alongside Pritchett, a clutch of business leaders used their control of a pension fund also funded by Carnegie to pressure the universities to adopt more businesslike management practices. Pritchett even turned to the famous efficiency expert Frederick S. Taylor for advice about how to accomplish this goal. One of Taylor's disciples wrote an extensive report recommending that professors become mere wage workers whose job would be to pump out as many student-hours as possible—with absolutely no allowance for the quality of the educational process.[39]

Similarly, in 1903, John D. Rockefeller founded the General Education Board, again run by corporate leaders. In 1910, the total endowments held by all the colleges and universities in the United States totaled just over $259 million; the General Education Board's endowment was $53 million in 1909.[40]

During World War II, the government dramatically stepped up its support of academic research. Vannevar Bush, the dean of the engineering school at MIT, went to Washington to head up the Office of Scientific Research and Development, which was responsible for the 6,000 scientists involved in the war effort.[41] In the process, Bush created what later came to be known as the military-industrial complex. His former employer, MIT, became the largest recipient of these research contracts.[42] In addition, Raytheon Corporation, a business that Bush cofounded, saw its sales increase 60-fold during the war, largely due to military contracts.[43] After the war, new government agencies, such as the National Science Foundation, for which Bush was the driving force, provided a new flood of research contracts to the academic world, including MIT.[44] As a token of the school's gratitude, it appointed him honorary chair of its board of directors.[45]

Ideally, academic institutions would offer researchers space for the disinterested search for the truth, and students the opportunity to develop critical reasoning skills and a broader acquaintance with the world. No university ever came close to meeting this ideal, but at some times and in some places the university environment came reasonably near. Of course, these instances were the exception, but they were important nonetheless.

Despite the strong commercial and military ties, much scientific research after World War II remained relatively independent. The assessment of Richard Nelson and Nathan Rosenberg, written in 1994, was applicable to earlier decades as well:

> In fact, the preponderance of university research today is in fields that, by their nature, are oriented toward facilitating practical problem solving in health, agriculture, defense, and various areas of civil industrial technology. On the other hand, the large fraction of university research that is classified as basic does indicate a certain distancing of much university research from immediate "hands-on practical problem solving."[46]

So while the majority of academic research, even at the elite institutions, was never pure science, the universities provided a substantial amount of pure research. This freedom allowed scientists to explore basic research without any perceived commercial value. Although this sort of research had little direct payoff, it eventually made vital contributions to the body of knowledge that made possible the technologies upon which modern society depends.

In recent years, the universities have been reinventing themselves to become more like corporations in their way of doing science cum business. The once-proud academic research infrastructure that had provided a hospitable environment for the development of pure science is fast disappearing. As public funding for the universities declined, they began to offer their services to corporate funders even more enthusiastically than before. In addition, universities are becoming major players in commercial fields, either collaborating or competing with the major corporations.

Changes in the legal system facilitated this transformation. The 1980 Bayh-Dole Act represented a key benchmark in opening up commercial possibilities. The idea sounded innocent enough: universities would be allowed to own the rights to research developed with federal funds. Prior to the passage of this law, universities were winning a total of only about 250 patents per year. Within a little more than a decade, the number of patents awarded to universities had risen to 1,600. Of those, nearly 80 percent stemmed from federally funded research. In addition, the number of universities participating in the patenting effort had increased to the point that in 1992, 200 universities had at least one patent issued annually.[47] While the idea behind this legislation seems innocent enough, in practice, the Bayh-Dole Act has meant that corporations on the "dole" would be able to "buy" universities.

Patents and royalties now loom larger for universities than the pride of contributing to the pool of human knowledge. For example, during fiscal year 1993, the top ten institutions in licensing royalties received a total of $170 million in royalty income.[48] The University of California alone received 395 patents in 1998,[49] and is even offering to sell or license some of its patents online.[50]

Patents and royalties do not yet represent a major source of revenue for universities, but with public funding in decline, academic institutions are counting on massive increases in revenue from such activity. One study of the subject observed, "We are witnessing the growth of universities as venture capitalists."[51]

Defunding Academic Science

This commercialization was no accident. Science requires support. At the time that the scientific societies that Paul David described emerged, scientists depended upon wealthy patrons. Later, universities offered support, and still later government agencies. This patronage became especially generous following World War II, when the prestige of science was soaring.[52] Even a few farsighted corporations generously funded pure science, although the most notable of these came from industries where serious competition was virtually nonexistent. I am thinking in particular about the stunning contributions coming from Bell Labs and IBM. Most scientific support, however, came from government rather than corporate sources.

The work of academic scientists sometimes seemed esoteric, unrelated to any possible social need. For example, Senator William Proxmire of Wisconsin, for the most part a sensible politician, used to ridicule some of this research with his well-publicized "Golden Fleece" awards, suggesting that the scientists and their universities were bilking the public by foolishly squandering money on ridiculous projects with no redeeming value. Proxmire failed to understand the way that science works. Often times, such seemingly frivolous scientific endeavors lead to unexpected breakthroughs in unrelated disciplines. In short, the calculus of politics or business profits provides a poor guide for the scientific process.

I am reminded about an earlier exchange regarding the usefulness of science. According to an oft-told but possibly apocryphal story, more than a century ago the famed British scientist Michael Faraday was trying to explain the discovery of electricity to Prime Minister William Gladstone. Gladstone's only comment was, "But after all, what use is it?" "Why, Sir," replied Faraday, "there is every possibility that you will soon be able to tax it."[53] Ironically, Faraday's portrait decorates Britain's 20-pound note.

Unfortunately, the heirs to Proxmire's anti-intellectual legacy eventually carried the day. Following the banner of the Reagan revolution, the enthusiasm for slashing government spending extended to undercutting support for higher education. Also, because the universities appeared to the leaders of this political movement to be far more radical than they really were, conservatives relished the opportunity to cut this seeming hotbed of antibusiness activism down to size. Moreover, by restricting funding for the universities, additional money could be freed up for tax cuts. Finally, an academic system starved of funds would have little choice but to turn to business interests to obtain financing.

Today, as a result of this strategy, public universities receive a rapidly declining portion of their resources from their respective state governments. At the University of California, Berkeley, for example, only 34 percent of the operating budget comes from the State of California, down over 50 percent from slightly more than a decade ago and nearly 70 percent from when Clark Kerr was president during the 1960s. For institutions with large medical centers, such as the University of Michigan, the percentage of operating budget from the state is only 11 percent.[54]

As funds got tighter, universities began spending significant amounts of money for lobbying to get more of the shrinking supply of funds. In 1999, Boston University spent more than three-quarters of a million dollars for lobbying. In addition, many universities fund offices in the capital to win more influence.[55] Although such activities may bring more money to a particular campus, they are unlikely to increase the funds available for academic science as a whole; they just deplete the already declining pool of available resources even further.

In this environment, public science, outside of military research, is withering from a lack of public funding. But military research is public only insofar as its funding comes from public sources. For the most part, it is shrouded in secrecy.

Starved of public funds, universities became more entrepreneurial. For example, in 1998, Berkeley negotiated an agreement between its Department of Plant and Microbial Biology and the Novartis Agricultural Discovery Institute. Novartis will provide long-term monetary support worth $25 million for basic research and, in exchange, receive first rights to license some of the discoveries made. In addition, the company will actually have some of its personnel sit on academic committees. Presumably, they will be able to influence what will and will not be researched. Considering the stakes, $25 million is not a trivial sum, but in an age when a single patent can be worth many times that amount, $25 million spread over a number of years is not a particularly generous contribution.

Similarly, the famed Oxford University made a deal with a British investment bank, Beeson Gregory. The bank will pay for one-third of a state-of-the-art chemistry building in the heart of Oxford in return for a share of the profits from any spinoff companies in the next 15 years. The bank's press release claimed that Oxford's chemistry department is the largest in the Western world, with 60 tenured faculty. Researchers, investors, management, and the university normally split equity in Oxford spinoffs—with the university typically getting 5 to 30 percent. Under this agreement, Beeson Gregory will get half the university's share in any spinoffs.[56]

The effort to serve the corporate sector can reach absurd levels. Consider the case of Petr Taborsky, an undergraduate college student in chemistry and biology, who took a job as a laboratory assistant at the University of South Florida College of Engineering in 1987. The lab employed him to do testing for a project studying methods to make sewage treatment cheaper and more efficient.[57]

On his own, Mr. Taborsky discovered a way to turn a claylike compound similar to cat litter into a reusable cleanser of sewage, a process that has many potentially valuable applications. He said that he made his discovery after the project had ended and that he did not conduct any of his experiments as part of his job.

The project's principal investigator, Robert P. Carnahan, maintained that Mr. Taborsky was part of a research team and that the discovery stemmed from the team's decisions. The university said the sponsor of the project, a subsidiary of Florida Progress, a utility holding company, had all rights to the research.

A jury convicted Mr. Taborsky of grand theft of trade secrets in 1990. He was sentenced to a year's house arrest, a suspended prison term of 3½ years, probation for 11½ years, and 500 hours of community service. Mr. Taborsky violated the terms of his sentence when he obtained three patents related to the research. He was assigned to chain-gang duty for two months, although he was later transferred to a work-release center in Tampa.

The Scandal of Scientific Journals

Defunding colleges and universities had other detrimental consequences. Traditionally, universities modestly subsidized scholarly journals as part of their commitment to furthering the overall educational process. With the drastic cuts in state support for education, financially squeezed universities cut these subsidies. No longer solvent, journals fell into the hands of great conglomerates.

Suddenly, these journals became a major source of profit. Prices skyrocketed. A subscription to an academic journal now can run to as much as $17,000, as is the case with *Brain Research*. These prices seem to bear no relationship to the cost of production. Reviewers, and even editors, frequently work without pay. Authors submit their articles in digital form, saving the publisher the cost of typesetting. In addition, journals frequently require a fee from would-be authors just for the privilege of having their work considered for publication.

A study of six academic journals found that the average journal costs $70,000 per year to produce, less than five times as much as a single subscription to Elsevier's *Brain Research*.[58] This comparison might be unfair, since that particular journal publishes about five times as many issues as a typical journal. Nonetheless, the profit from this business can be spectacular. The business press reports that the publisher Elsevier, which also owns Lexis-Nexis, makes about a 40 percent rate of return on its scientific publications.[59]

Furthermore, the conglomerates should be able to produce these journals more efficiently than nonprofit universities and professional societies had. To begin with, we should not expect to find great managerial efficiency in the nonprofit institutions. In addition, the conglomerates could enjoy economies of scale.

For example, in 1975, a study found that the cost of producing a scientific journal by a company already producing 30 journals is 80 percent that of a company producing a single journal. Consequently, the great conglomerates should have been able to lower costs.[60] In 1991, when Elsevier was about to buy Maxwell's scientific publishing unit, Elsevier was already publishing about 650 journals a year.

With soaring journal costs, the typical university library can afford fewer and fewer journals—especially those that cost more than $1,000 per year. Rising journal costs also force libraries to scrimp on the purchase of books. According to Walter Lippincott, director of the Princeton University Press, "In 1975 a library might spend 70 percent of its annual budget on books and 30 percent on journals. Now those percentages are reversed."[61]

Certainly, this restriction on the supply of information seems inconsistent with the concept of an information age. For the society as a whole, the previous cost of subsidizing professional journals was relatively trivial. After all, if less-restrictive access to scientific information would allow for a single significant scientific discovery, this outcome would pay for the subsidies many times over.

In the Third World, the cost of scientific journals is even more prohibitive. Some entire countries can afford only a handful of publications. Researchers in the Third World often cannot publish in scientific journals, because they cannot afford the publication fees that journals charge.[62] As a result, large parts of the world are virtually cut off from much scientific discourse. We all suffer from the lost opportunity to benefit from the full intellectual potential of these scientists, but, alas, science must give way to the dictates of the marketplace.

In 1988, when the problem of academic journals was less severe, Henry H. Barschall, a retired University of Wisconsin physicist, published a comparative analysis of more than two hundred publications. Not surprisingly, he found that nonprofit society-based journals offered work equal to or better than commercial journals, and at lower subscription prices. Gordon & Breach, a commercial publisher whose journals came out poorly in the study, demanded an immediate and prominent retraction of the article.

The nonprofit publisher of the journal in which the article appeared offered Gordon & Breach the opportunity to publish their objections, but intended to give Barschall a chance to examine the corporation's submission and counter any statements of error—standard practice in scientific journals. Gordon & Breach sued the American Physical Society and the American Institute of Physics. After $3 million in legal costs, a judge ruled in 1997 that the suit was but one battle in a "global campaign by Gordon & Breach to suppress all adverse comment upon its journals."[63]

Aping the Corporations

Even public funds for research have become increasingly tied to corporate objectives. Business leaders have successfully pressured the federal government to sponsor commercial research and development in research universities and in government laboratories. The National Science Foundation, once regarded as the bastion of basic research, began developing industry-university cooperative research centers in the 1980s. The Department of Commerce now pursues a national science and technology policy exemplified by the Advanced Technology Programs, begun under the Clinton administration.[64]

As a result, academic research, as well as some government research, has ceased to be directed toward contributing to the public domain. Profit-oriented universities have severely cut back on basic science that might make future technological breakthroughs possible; instead, they are putting a premium on work that is likely to promote corporate funding or quickly lead to income from intellectual property rights.

In 1964, Mario Savio galvanized students around the country with an impromptu speech from atop a police car that helped to ignite the Free Speech Movement at the University of California, Berkeley. Perhaps the high point of his oration came when he announced: "if this is a firm, and if the Board of Regents are the board of directors, and if President Kerr in fact is the manager, then I'll tell you something: the faculty are a bunch of employees, and we're the raw material!"[65] Since then the rot of entrepreneurialism has worked its way even more deeply into the structure of the university than Mario Savio could have imagined.

Universities are beginning to hire more and more administrators from the business world rather than from within academia. Even those administrators who came up through the ranks of academia have begun to fancy themselves to be CEOs. In keeping with their entrepreneurial pretensions, many universities have proven to be as ruthless as private employers in their efforts to oppose unions and to keep wages low.

In the process, the spread of business jargon has become commonplace among administrators. For example, students have become "clients." Reflecting the recent business craze for downsizing, President Richard Levin of Yale University, one of the earlier-cited researchers who studied the costs of circumventing patents, unveiled the principle of "selective excellence" in 1996. According to Levin, "no university . . . has the resources to be the best in the world in every area of study." Therefore, he explained, "our programs should be shaped more by an aspiration to excellence than a compulsion to comprehensiveness."[66] This call for downsizing seemed to imply that universities should abandon low-value-added activities. Soon thereafter, university presidents around the country began to cite Levin to justify their own downsizing agendas.

No university president would dare to identify teaching as a low-value-added activity, not even the teaching of the humanities, but increasingly universities have allowed their teaching mission to fall into the background. The constant search for money affects virtually every aspect of university life. I teach at a university that began as a teachers college. Historically, professors were expected to teach and do little else. Today, the university frequently expects new teachers to give evidence of their ability to win grants as a precondition of their employment. Since the supply of grants is relatively limited, faculty members dissipate increasing amounts of time in competing against each other for the same pool of grants, often to the detriment of their teaching. It goes without saying that this criterion narrows the pool of prospective teachers to those who would be uncritical of the potential grant givers.

In this atmosphere of high-value-added education, what teaching that remains is increasingly seen as a service to business. Administrators "reform" the curriculum so that students concentrate on acquiring the skills that corporations expect of their employees. In effect, the university is increasingly becoming a trade school for those young people who can afford to attend.

All the while, administrators, in further emulating business practices, have been seeking concrete measures of accountability, even though the numbers that they are using to measure efficiency have no relevance whatsoever to the actual learning process.[67] While demanding accountability from others, university CEOs tend to spend lavishly on their own projects. They clamor to accumulate outward symbols of their university's achievements. In the process, they compete to build the most impressive structures and to hire the most prestigious faculty, as in the case of Robert Barro, whom I will discuss in the next section.

Such trophies are supposed to increase the prestige of the institution in the eyes of the corporations, which will view such enterprise as evidence of success. After all, the corporations are ill equipped to recognize what makes a fertile academic environment. A lavish building or a widely recognized name, however, may do much to signal to them where they should invest their money. In the end, though, bidding wars and lavish construction binges do relatively little for education. For the most part, they are a zero-sum game. After all, only one institution can be number one.

Playing the prestige game can run up enormous costs. Administrators often make up their budgetary shortfalls by cutting back in other areas. Consider the City Colleges of Chicago, which with more than 160,000 students represents a harbinger of the future. In early 2001, the system began turning over control of its $284 million annual operating budget to American Express Tax and Business Services Inc. The college presidents have been asked to submit reports evaluating business and computer science instruction, plus library and counseling services.[68] Major universities, such as Harvard, routinely contract out nonacademic jobs to other companies, which can achieve greater "efficiencies" by cutting back on salaries and benefits, such as health care.

A cynic might respond that the welfare of janitors and food service workers should not rank high in the priorities of an institution of higher learning. Instead, universities should muster all the resources they can to fulfill their mission.

In their zealous pursuit of corporate emulation, some universities may be inclined to abuse the intellectual property of their researchers. In several cases, scientists have sued universities for ignoring their interests in granting licenses to companies. These deals shortchange the researchers in return for a promise of corporate-sponsored research at the school.[69]

The most common strategy for making up for administrative excesses is to raise tuition.[70] Everywhere tuition is soaring. Tuition, books, and room and board at an elite university total about $35,000 per year. In short, making the university more entrepreneurial ends up making college education less affordable to all but the affluent. Students attending college now work more and more hours to keep up with the expenses.

The net effect is to reinforce the existing maldistribution of income. Poorer students often do not attend colleges, and if they do, they cannot learn as much because of excessive job commitments. Wealthier students attend elite universities, where the corporations go to recruit for the highest-paying jobs, again reinforcing the skewed distribution of income.

The Imperial University

The Massachusetts Institute of Technology exemplifies the changing nature of the university. The school has deservedly maintained an outstanding reputation for scientific excellence, despite its strong corporate ties. Since 1994, the university has set up eight long-term research partnerships with major corporations, including Ford, Du Pont, Microsoft, Merck, and Merrill Lynch, which will generate $248 million over the next 12 years.[71]

The university has also profited handsomely from its scientific output. Like few other elite universities, MIT has provided the seedbed for a number of high-tech companies. By 1993, the school was reported to have already spun off 836 businesses with sales of $60 billion.[72] No doubt this figure has multiplied many times since then. Frank Rhodes, Cornell University president emeritus, cites statistics that show that the alumni and faculty of MIT have spawned some 4,000 companies that employ 1.1 million people and generate $230 billion in annual sales—a feat that as a stand-alone economy would rank it twenty-third in the world, "between South Africa and Thailand."[73]

In return for its services to many of these companies, the school has enjoyed mutually beneficial relationships. Consider the example of Amar Bose, MIT professor and founder of sound-system maker Bose Corporation, a company now worth more than a half-billion dollars. In the early years of Bose Corporation, MIT allowed the company to use its labs rent free until it could find permanent offices. The school also invested in the company.

This relationship proved profitable to both parties. MIT shared in the prosperity of the company as it matured. In addition, the Bose Foundation has since donated more than $6 million to MIT. The university has accumulated a $4.3 billion endowment with the help of such largess.

Bose's son, Vanu, recently earned his Ph.D. from MIT. The university owns the patent on technology that he devised for his dissertation because that work was done at the university. Vanu wants MIT to give his start-up company, Vanu Inc., exclusive rights to the patent. In return, however, MIT demanded $1.25 million over the next eight years in licensing fees, along with royalties of 10 percent on licensed services, 10 percent on software the firm develops, 4 percent on computer hardware, and 6 percent on software embedded in hardware. In addition, MIT asked for 6 percent ownership of the company.[74] Although the university later moderated its demands, the outcome still remains in contention.

In an even more revealing case, a student at MIT was unable to complete a homework assignment. He knew the answers. He had another reason. To do the assignment would have violated the nondisclosure agreement that he had signed with a company run by another of his professors. The professor who gave the homework had organized his own company, which had objectives similar to the company with which the student had signed the nondisclosure agreement. The other professor was reported to have questioned whether the assignment was designed merely to give an indication of the student's knowledge or whether it was intended to elicit information about the competing company.[75]

Given this welter of conflicting interests—with students doubling as both employees and holders of stock options in their professors' companies and with professors acting as both teachers and entrepreneurs—traditional academic values are certain to recede. One MIT student registered his concerns about the effect of such entrepreneurial activity on teaching in a stinging commentary in the student newspaper. He complained that teaching suffers when professors are instructing a class while simultaneously recruiting students. Intent on discovering which people to hire, they modify their teaching methods. Rather than providing clear explanations of their material, these academic entrepreneurs try to overwhelm the class in an effort to identify the best possible employees. They then devote an excessive amount of attention to those pupils, to the detriment of the rest of the class.[76]

Stanford University has taken the entrepreneurial mission one step further. It has become the first university to launch its own for-profit business—an online medical search engine called e-Skolar Inc.[77] Columbia has entered into a partnership to create an online information company, NutritionU.com, with a company that sells nutritional supplements.[78] Other universities are organizing their own venture-capital funds.[79] Meanwhile, the Association of University Technology Managers frets that universities spent $24 billion in 1998 on research and development, but realized only $739 million in revenue—as if the main purpose of academic science were to turn a profit.[80]

If current trends continue, however, academic science will be earning a profit in the not-too-distant future. Columbia University's licensing revenues already account for almost 25 percent of its entire research budget.[81] These revenues were expected to increase by 44 percent in the year 2000. Keep in mind that research brings in revenue through other avenues. For example, corporations often contribute funds in return for collaboration on research.

Academic Hierarchy and Superstars

This madcap quest for corporate funds is rapidly deforming the academic community. In their enthusiasm to tap the commercial potential of their researchers, the elite universities are busy reconfiguring their academic programs—and not for the better. They recruit "star" professors with the same zeal, although for slightly less money than owners of professional athletic teams spend in their efforts to assemble talent. The objective is to make the institution more attractive to the corporate sector.

For example, Columbia University tried to lure Robert Barro, an economist from Harvard, to join its economics department for an annual salary of nearly

$300,000. The working conditions that the university promised were even more generous. Professor Barro was to occupy three spacious offices on the impossibly cramped and overcrowded campus. He was to direct a sizable chunk of resources at the university's new social science research center. Most important, Professor Barro was to have a green light from the administration to recruit a half-dozen promising younger economists.

The university also promised to make life better for the rest of his family. It helped to place his teenage son in an exclusive Manhattan private school. It offered a $55,000-a-year university post to his wife, Judy, who had been out of the work force. Finally, it eased out the tenant of a 2,300-square-foot university-owned apartment on Riverside Drive that the Barros coveted. The university also offered to subsidize the rent for this apartment. To make the offer even more attractive, it promised a lavish renovation. In the end, Professor Barro relented and remained at Harvard, presumably because Harvard must have improved his working conditions there.[82]

Robert Barro's offer may be atypical, because economics does not have as direct a tie to business as do biotechnology or computer science. In the future, offers such as this one will probably not be particularly unusual. With the commercialization of the university, academic employees are less likely to evaluate their situation by traditional academic standards. Instead, they are tending to compare their work with the monetary rewards that they could enjoy working for business or by starting their own companies. Already, some technical disciplines are suffering a painful brain drain.[83]

This bidding war is a zero-sum game. One school's victory is another school's loss. In the process, less and less money will be available for other uses. For example, universities are increasingly staffing their less commercially viable courses with underpaid part-time or adjunct instructors who often receive shabby treatment. Even less-established junior faculty are not immune from this fate.

For example, John McLaren, a young economist, had a less royal encounter with Columbia's economics department than that promised to Professor Barro. McLaren concluded that the department viewed junior professors as underlings "to be flattered and fleeced." By the time he left the department, he had never received an upgrade for the obsolete 486 computer he got when he arrived in 1992, although a modern computer is essential for managing economic data. The ceiling over his desk caved in, showering powdery debris over his obsolete computer and swinging a sharp swordlike metal rail down around where his forehead would have been if he had been at his desk at the time.[84]

If Professor McLaren were a failure or an incompetent, his complaints might not seem to be worth taking seriously. In fact, even though the department kept him working on time-consuming administrative duties and had him prepare a greater variety of courses than other professors, including a demanding introductory course, he managed to publish ten papers in prestigious economics journals, enough to get offers of tenure from the University of Virginia, Boston College, and the University of Wisconsin.

Besides the funds for joint ventures with university researchers, the corporate sector helps to finance the construction of lavish buildings for those departments with the most direct ties to business interests. Universities' neglect of the infrastructure for academic disciplines less favored by the corporations is appalling. Those facilities have come to resemble academic slums. I will return to this subject later.

The Faustian Bargain

I mentioned earlier that universities earn sizable revenues from their scientific research over and above what they earn from patents and royalties. Because corporate science is less efficient than traditional science, corporations gladly turn to the financially strapped university laboratories with a Faustian bargain: Hitch your labs to our budgets and we will ensure you of adequate funding, but you must accept, to a greater or lesser extent, our objectives and our needs. You must abide by our demands for secrecy and you cannot raise questions about any untoward effect of the products that we intend to produce.

As universities, especially the more prestigious ones, become beholden to corporate powers for funding, the corporations reap a threefold windfall. First, university research is cheaper than corporate research. The universities supply buildings, libraries, and other facilities, as well as low-wage graduate students. More important, other things being equal, scientists prefer working in an academic atmosphere.[85] As a result, universities produce better research than corporate laboratories. Second, corporations can project a philanthropic image by funding universities. Finally, and perhaps most important of all, the corporations get to set the national research agenda.[86]

Given university dependence on corporate funding, the corporate sector can silence potential critics, or at least shroud their work from the view of potential critics. After all, anybody within the university who questions the corporate agenda poses a risk to the university's finances.[87]

Just imagine what the status of scientific knowledge about the effects of smoking would be in a world dominated by corporate science. Suppose that research on tobacco were to begin today from square one, with no prior knowledge about the health effects of tobacco. What corporate sources would be willing to support research into the dangers of smoking? Quite likely the most willing benefactors of tobacco research would be the tobacco companies themselves, with their obvious vested interest in the outcome. Universities might be leery of hiring a researcher with expertise or even interest in the detrimental health effects of smoking if they wanted to lure a tobacco firm to subsidize their research. Under such conditions, the deleterious effects of smoking would be almost impossible to establish.

This danger is not purely hypothetical. Take an example of an intellectual property owner from outside the realm of science and technology. Naomi Klein reported:

> In May 1996, students and faculty at the University of Wisconsin at Madison did find out what was in the text of a sponsorship deal their administration was about to sign with Reebok—and they didn't like what they discovered. The deal contained a "non-disparagement" clause that prohibited members of the university community from criticizing the athletic gear company. The clause stated: "During and for a reasonable time after the term, the University will not issue any official statement that disparages Reebok. Additionally, the University will promptly take all reasonable steps to address any remark by any University employee, agent or representative, including a Coach, that disparages Reebok, Reebok's products or the advertising agency or others connected with Reebok."[88]

Would sponsors of scientific research be more understanding? Paul and Anne Ehrlich, pioneers in raising environmental consciousness during the late twentieth century, described how a handful of well-funded scientists and journalists disputed the dangers of environmental damage from global warming and pollution.[89] Fortunately, those who toe the corporate line in meteorology are still in a minority, probably because the majority of leaders in the field matured in an environment in which universities still provided some refuge from commercial pressures.

At Nottingham University in Britain, British American Tobacco, a company that sells a lethal product and that was under investigation for smuggling at the time, has set up a center for corporate responsibility.[90] At Tulane University, Freeport McMoRan, a mining company with operations around the world, has established a chair of environmental studies.[91] Environmentalists have singled out Freeport McMoRan because of the widespread devastation associated with its mining practices. Will the person holding this chair be willing to take an impartial stance regarding Freeport McMoRan? Will the university be willing to appoint a person whom they expect could possibly be critical of the company?

Indeed, a number of cases suggest that universities tend to refuse employment to researchers who might be critical of a major funder. In one case documented by the Toronto *Globe and Mail,* the University of Toronto even rescinded a job offer to a scientist, Dr. David Healy, whom they had been courting since July 1999. The university made him a formal written offer of a combined faculty and clinical position in May 2000, followed by a more detailed letter in August. They even hired a lawyer to help him immigrate.[92]

Then on November 30, 2000, Dr. Healy gave a wide-ranging lecture at the university's Centre for Addiction and Mental Health. He criticized pharmaceutical companies for avoiding experiments that could demonstrate problems with their drugs and for not publishing unfavorable results. He said the data show that Prozac and other popular antidepressants in the same chemical family may have been responsible for one suicide for every day they have been on the market.[93]

A week later, Dr. David Goldbloom, physician-in-chief at the Centre and a professor at the university, rescinded the offer to Dr. Healy in an email. The Centre's Web site indicates that its "lead" donor was Eli Lilly, the same drug company that manufactures Prozac. Similarly, in 1999, Eli Lilly canceled its $25,000 annual donation to the Hastings Center in New York, a think tank that looks at ethical issues, after the latter published a series of articles about Prozac, including a critical one by Dr. Healy.[94]

The stifling of critical academic research is more dangerous than ever before, now that government agencies see the corporate sector as their clients. With minimal regulatory oversight, limiting independent criticism pretty much gives the corporate sector free rein.

The Deeper Surrender of the Universities

In general, while no expense is spared to build up the prestige of their commercially oriented research arms, at the other end of the spectrum, the modern universities typically let the humanities wither. One commentator has observed the enormous gap that is opening up between the favored business-friendly disciplines and the more traditional academic pursuits:

> [L]ess lucrative academic pursuits are maintained as museums of the outmoded, and their curatorial faculty is paid a comparative pittance. As recently as a quarter-century ago, approximate parity in salaries was the norm, but today

> a full professor of English earns no more than a starting assistant professor of accounting.[95]

Libraries at the major universities are excellent indicators of the nature of the entrepreneurially oriented academic universe. For example, at the University of California, Berkeley, the libraries of the School of Business and Department of Biological Sciences offer comfortable, even plush environments for doing research. Their collections are enormous. The Business and Economics library provides numerous outlets that allow a laptop computer to connect to the Internet. At some of the libraries in the less commercially connected branches of the university, the environment seems positively impoverished by comparison. Hardly anything has changed since I studied there in the 1960s, except for the aging of the facilities.

The commercial fervor of the universities more or less obliterates any distinction between corporate and university research. Derek Bok, former president of Harvard, in his final report to the university's Board of Overseers, found "the commercialization of universities as (perhaps) the most severe threat facing higher education." He went on to warn that universities "appear less and less as a charitable institution seeking truth and serving students and more and more as a huge commercial operation that differs from corporations only because there are no shareholders and no dividends."[96]

According to the University of Miami's vice-president for research, "As money becomes less and less available, more people are going to be compromising their principles, compromising their time. . . . We can get to the point at some stage in this process where we're not research universities any longer but fee-for-service corporations—hired guns."[97]

In this world, academic careers rest on the ability to land corporate or government (mostly military) contracts. Researchers can either work at the behest of corporate "donors" or attempt to become independent by seeking out profitable discoveries either to patent or to use as the basis for their own firms.

The legal system is bending over backward to accommodate such practices. Today, when a biologist can patent a sequence of genetic material or a mathematician can patent an algorithm, money rather than the acclaim of colleagues becomes the coin of the realm. Researchers, who once worked in the open to win recognition from their peers, now shroud their research in secrecy in the hope of striking it rich.

In effect, then, this epidemic of entrepreneurialism, infecting both the universities and a good number of their researchers, contaminates the scientific process with the same sort of defects that impede corporate science. Because so many leaders have deluded themselves into thinking that the recent successes in both science and the economy rest wholly on the individual efforts of private enterprise, they feel confident in hobbling the very institutions responsible for so much technological progress.

This strategy is likely to turn out a sizable number of clever innovations in the short run, but it will certainly undermine the basic scientific process upon which innovations in the more distant future will depend.

Secrecy vs. Science

Of course, if this new system of profit-oriented research were to provide effective incentives that somehow made possible a superior technology that would improve the lives of a significant share of the population, then the erosion of intellectual freedom within academia might be a small price to pay. Unfortunately, I doubt that science oriented toward intellectual property will prove to be particularly effective, for several reasons.

In the first place, as scientists and universities become profit minded, scientific behavior changes, especially within a regime of strong intellectual property rights. Intellectual property rights may well encourage an individual scientist or even an individual firm to redouble efforts in a particular line of research, but the lure of potential profits from potential intellectual property rights may very well harm the sort of open communication upon which a healthy scientific process depends.

Since commercial considerations must remain uppermost in this environment, the valuable tradition of sharing information, which has made modern science so effective, becomes shrouded within a veil of secrecy. In this environment, scientists no longer can afford to engage in so much open communication with their counterparts. Instead, fellow scientists become rivals in the quest for profits rather than colleagues in the search for the truth. The free flow of information, which was once a hallmark of science, turns into a threat to potentially profitable opportunities.

So, although the quest for profit might make an individual researcher work harder, the establishment of intellectual property rights will probably harm the scientific process as a whole, by inhibiting the network effects. As a result, looking at the actions of an individual researcher might give the misleading impression that intellectual property rights encourage science, while, in fact, such is probably not the case.

Consider the planning of science in the corporations that produce pharmaceuticals. The scientists who investigate the properties of a potential pharmaceutical product often discover that it serves a completely unexpected use. Since labs in the larger firms have more people working on a wider range of projects, they are more likely to find a potential use for any particular chemical. As a result, in this industry the larger firms tend to be more productive.[98]

Now imagine that the large pharmaceutical company is an amalgam of many separate, small companies. In the absence of their common corporate employment, many of the researchers would not have the opportunity to communicate with each other about their work. Within the corporation, this larger group of researchers can communicate, since the corporation can reap the rewards of what one scientist learns from another. In this sense, the communication among scientists within the large firm may resemble on a small scale what science was like in a more open world, but with obvious differences.

In an open world, the cross-fertilization within the large pharmaceutical company would be part of a larger process whereby progress in one industry sparks important innovations in unrelated industries. These potential spillovers imply that the more scientists are free to cooperate and collaborate, unrestricted by the need for corporate secrecy, the more society will harvest from the scientific process. Thus, the fragmenting of science into small enclaves ruled by independent corporate powers discourages scientific exchange.

Rather than promoting the free exchange of ideas the corporate culture in which ideas become private property puts a premium on secrecy. For example, an editorial in *The Lancet* reported that when Martin Cormican, a bacteriologist at University College Hospital, Ireland, wrote to Bayer in November 2000 asking for a supply of pure ciprofioxacin and related products for his research into antibiotic resistance, he was asked to sign a document, stating, "We declare that we will inform Bayer AG in writing of our test results and will not publish or commercialise them without written permission of Bayer AG."[99]

Efforts to maintain secrecy impose direct costs, over and above those that detract from the scientific process. Consider, for example, the elaborate system of envelope security practiced at IBM:

> The envelope hierarchy went like this: IBM Internal Use Only; IBM Confidential; IBM Confidential Restricted; and Registered IBM Confidential (the top secret category). There were two sizes of envelopes for each level of security. IBM bought about four million envelopes a year. A rulebook spelled out which level of bureaucracy had access to each level of confidentiality. But IBM ended up branding hundreds of mundane documents IBM Confidential, even though they weren't. Even assuming IBM was able to buy these specially printed envelopes at the bargain basement price of 10 cents each, it was still spending $400,000 a year on envelopes alone. The costs of protecting computer-transmitted information may be very much greater.[100]

Worse yet, these costs will certainly escalate in the future. While modern technology makes the spread of information less costly, it also makes the maintenance of secrecy more difficult. Ironically, the companies that are so determined to maintain their secrecy avidly collect and use personal information about their customers in ways that violate what most people would consider to be norms of privacy.

I suspect that one reason that economists have not yet addressed the full implications of the privatization of science is that economics habitually frames processes in terms of individual agents—a firm, a consumer, etc. Scientific progress is rarely the product of an individual researcher. Instead, science is a network process. It advances through cross-pollination. Scientists learn from other scientists. In addition, they take information and inspiration from people in fields outside of their own, as well as from close colleagues.

For example, in 1999, a number of people in New York were afflicted with the virus, initially diagnosed as St. Louis fever, that is spread by the Culex pipiens mosquito. Fortunately, Tracey McNamara, a veterinarian and pathologist at the Bronx Zoo, was called upon to study the widespread deaths of crows in the Bronx and Westchester County. Soon afterward, five Chilean flamingos turned up dead, along with a cormorant, a snowy owl, and an American bald eagle. She diagnosed the cause of the birds' demise as West Nile fever, a rare disease previously unknown in the United States. Eventually, the Centers for Disease Control and Prevention was alerted to the true cause of the human illnesses. As a result, the public health authorities were better prepared to protect the population.[101]

A second case, regarding pharmaceuticals, may be even more telling. When people ingest medicines, frequently they excrete the active agents without breaking them down. More often than not, these chemicals move through the sewage system and into the waterways. As a result, substantial amounts of chemicals are concentrating in the water and also on land where sewage sludge is applied.

For the most part, nobody has examined these drugs for adverse environmental effects, even though many of them "are designed to modulate endocrine and immune systems and cellular signal transduction and as such . . . have obvious potential as endocrine disruptors in the environment."[102] Many of these chemicals are designed to have profound physiologic effects on humans; they would be likely to affect nontarget organisms as well. "A major unaddressed issue regarding human health is the long-term effects of ingesting via potable waters very low subtherapeutic doses of numerous pharmaceuticals multiple times a day for many decades."[103] Maybe the problem of pharmaceutical pollution will turn out to be inconsequential, but it does raise what would seem to be a serious question.

In a scientific world of specialization, where everybody is intent on contributing to the commercialization of some product, who would ask such questions? What organization would promote the sort of communication that would allow people who understand medicine, biology, hydrology, sanitary engineering, and other disciplines to come together and figure out the extent of the problem?

In conclusion, science is a process rather than a simple act, such as buying an apple. The secrecy associated with intellectual property is antithetical to this networking. As I will discuss later, economics, because of its emphasis on the individual transaction, is ill suited to modeling something as complex as the vigor of the network process.

The Stifling of Science

Even where corporations attempt to impose secrecy, the free movement of people from company to company can overcome such restrictions, thereby spreading vital information. Of course, this behavior also violates the principles of intellectual property rights. In fact, a number of astute observers have perhaps correctly attributed the achievements associated with Silicon Valley to a mobility that defeats, at least in part, the notion of intellectual property rights. Specifically, they point to the ease with which employees in the region move from firm to firm, taking their knowledge, including corporate secrets, with them. In contrast, the atmosphere along Route 128 near Boston precludes people from enjoying the same degree of mobility.[104] More recently, the Silicon Valley's success in computers has repeated itself in networking technology. Because of a stronger climate of openness, the technology emerging from Silicon Valley has remained on the cutting edge, while firms around Route 128 have more frequently faltered.[105]

The openness that people have attributed to the environment of Silicon Valley may be a thing of the past. The courts are now modifying the legal structure that permitted the sort of mobility that once supposedly contributed to the success of Silicon Valley.[106] Even more dramatically, the people of the region themselves seem to be thoroughly embracing the culture of secrecy. For example, people now routinely require others—even loved ones—to sign legally binding nondisclosure forms before engaging in any conversation that has to do with work.[107] This fear of disclosing secrets is certain to put a chill on the sort of open communication characteristic of a successful scientific community.

Over and above the problems created by an atmosphere of secrecy, markets are an inappropriate venue for setting scientific priorities. Business is reluctant to fund projects that do not promise to deliver relatively high profits in a short period of time. Recall the 20-to-30-year time lag between a scientific discovery and its eventual appearance in commercial products. This lag might be shrinking because of business's emphasis on rapid commercialization, but it remains significant nonetheless.

In addition, business generally makes plans that tightly focus on a narrow commercial objective. Basic scientific research cannot focus in the same way that a business can. As noted previously, many of the greatest scientific discoveries are serendipitous. A scientific investigation usually begins by looking into a specific area of interest, often within a very narrow framework. Nonetheless, the results frequently lead into unexpected directions, providing vital leads for researchers in very different fields. Even though the ultimate result may pay huge dividends in terms of social benefits, business cannot plan for such an outcome. As a result, funding basic research is an unattractive proposition for business.

Even if business could somehow anticipate the outcome of a scientific venture, it still might have little incentive to fund such research. Most important scientific endeavors have a long gestation period. Since the research may not pay off for a generation, business, with its eye ever on the next quarterly report, would be reluctant to make such an investment.

Even if firms were willing to take a long-run perspective, the profits required to justify basic research are enormous. For example, the pharmaceutical industry makes about 15 percent per year on invested capital. Presumably, the industry would resist investments that make less profit.

Recall the earlier estimates that basic scientific research takes about 20 years before it comes to fruition in the form of a commercial product. Assume that a company could be absolutely certain about the outcome. Compounding 15 percent rate of profit over 20 years would mean that each dollar spent in basic research today would have to promise at least $15 twenty years from now to justify the investment.

Of course, the company has no way of knowing the future payoff of any research effort. Because of the pervasive uncertainty surrounding the consequences of basic research, the payoff would have to be many times more than the $15 to justify the investment. In short, competitive business is reluctant to engage in basic research.

Over and above these problems, business considerations stifle scientific progress because commercial pressures divert a significant amount of attention from doing research. Susan Zolla-Pazner, a professor of pathology at New York University, editorialized about this problem in a back-page article in the *Scientific American:* "when today's professors hit the big time, they have to read their professional literature and *Business Week,* write scientific papers and patent applications, teach, give seminars, and sit on the scientific advisory boards of various corporations."[108] She continued:

> The academic scientist finds herself taking a crash course in business and law. The demands of negotiating agreements and writing patents drain time and energy. . . . The basic research that sparked the initial effort may lie fallow. The spontaneity of scientific pursuit, so prized by those lucky enough to have investigator-initiated government research grants, may be restricted. The speed with which the professor can share data or new reagents may be slowed. The result, in the worst scenario would be deleterious for the lab, harmful for science, bad for society.[109]

These observations are not unique. According to the director of research and development of one of the largest biotechnology corporations, "the most important publications for our researchers are not the chemistry journals but the patent office journals around the world."[110] I will discuss later how considerations of intellectual property shape science in the laboratory. Scientists often find themselves compelled to choose inferior strategies either to develop stronger claims to intellectual property rights or to avoid having to pay royalties to others.

The Stakes of Scientific Research

Here we come to still another problem with commercially oriented science. I have already noted that when a scientist makes a significant contribution that leads to a commercially viable discovery, the ultimate beneficiary of the discovery may be another firm or even another industry. The employer of the scientist may have difficulty in collecting royalties from other firms that have benefited from the discovery. Even if the firm manages to collect the royalties, it may have to dissipate huge amounts of resources in legal wrangling before it succeeds.

Since those who fund science cannot be assured that they will be able to reap all the rewards that flow from their laboratories, business will invest less in science than would otherwise be the case if society as a whole were to make the decision. Economists have long understood that markets lead to less investment in science than would be socially desirable, but they have never been able to devise a method to avoid this problem within the context of a free-market economy—even with a strong patent system.

A strong system of intellectual property rights creates its own problems. I have already discussed a number of the destructive effects of intellectual property rights, including wasteful litigation, duplication of effort, loss of information due to secrecy, and a commercial atmosphere that depresses creativity. Let me add another destructive effect here: the royalties charged by one firm become a cost to another. Since these royalty costs make research more expensive, they discourage firms from engaging in research and development.

With less investment in science, society will fall short of its technological potential. This danger becomes increasingly threatening because of the growing dependence upon technology. As the world becomes more complex and an expanding population attempts to pump a greater standard of living out of the environment, the world will demand even more complex technologies based on new scientific discoveries.

Think for a moment about the discussions leading up to the year 2000 problem that was supposed to disrupt computers around the world. Even though the dire catastrophes that some predicted did not materialize, the warnings about this impending disaster should have reminded everybody about how much of our daily lives rely on technologies that we take for granted. Even more important, the more that modern societies attempt to control the environment or to overcome environmental limits, the more precarious conditions become.

Just consider the role of modern medicine. Ease of travel combined with densely populated pockets of intense poverty is certain to spread disease. Global warming is already moving tropical diseases into what are considered to be temperate parts of the world. Chemical agents that upset the immune system are spread with abandon. People rarely consider this dependence on science until some threat, such as the AIDS virus, makes itself known. Nobody can anticipate the exact form that these threats may take. When they do occur, science must be prepared to meet the challenge by responding to the escalating demands for a techno-fix.

Will corporate science be able to meet that challenge? To date, it has not offered much evidence that it can.

Universities and Secrecy

Many of the conditions that made university research superior to corporate research are fast disappearing as the universities bow to corporate pressure at the same time that they are beginning to emulate the corporations.

Within the traditional university environment, science was a way of life. Scientists did not expect to become fabulously rich from their discoveries. Instead, as noted earlier, they vied with each other for the honor and prestige of contributing to the promotion of scientific knowledge. Sharing was an integral part of the process.

The military assembled some of the most important researchers at the RAND Corporation in the period following World War II. While secrecy in relation to the outside world was imperative, the people who ran the organization understood the importance of sharing information, even among people from different disciplines. They devised an architecture that would maximize the frequency of chance encounters.[111]

Derry Roopenian, a biologist at the Jackson Laboratory in Bar Harbor, Maine, reminisced about the time when sharing ideas was highly valued: "At one time, if you found something exciting, you would run down the corridor and talk about it. . . . Now if you discover something big a commercial backer is interested in it, you can't say a word about it."[112] Other conversations would seem to have nothing whatsoever to do with their work. For example, Francis Crick, who won the Nobel Prize in medicine in 1962 for discovering the structure of DNA, shared an office with Sydney Brenner for 20 years. They had a rule of uttering "anything that came into your head." According to Brenner, much of their conversation was "just complete nonsense [but] . . . every now and then a half-formed idea could be taken up by the other and really refined."[113]

When a single offhand remark sparks a new line of scientific research, the dividends for society can be incalculable. In the world of corporate science, however, other values take precedence. Research has to be directed to shorter-term projects with obvious commercial value. Communication must be censored to keep profitable ideas from falling into the wrong hands.

The same sort of limits on communication that the corporate world dictates is becoming common within the university research systems. For example, corporate sponsors of research commonly impose a number of conditions on their grants. To begin with, corporations often impose conditions on university researchers that allow the firm to squelch any research that leads to unwelcome conclusions. In one notorious example, Boots Pharmaceuticals gave Betty Dong of the University of California at San Francisco $250,000 to study Synthroid, a synthetic form of thyroid hormone taken daily by eight million people at an annual cost of some $600 million. Boots asked Dong to determine whether three generic forms of the drug were biologically equivalent to Synthroid, presumably expecting the answer to be negative. When Dong discovered that the generics were equivalent and tried to publish her results, Boots invoked a clause in the research contract to force Dong to withdraw the paper. Here the matter might have died had the *Wall Street Journal* not uncovered the episode.

In the glare of publicity, Boots finally backed off. The paper was finally published in April 1997, two years behind schedule.[114] The cost of secrecy in such matters can be substantial. For example, a few years later, the Food and Drug Administration (FDA) told the makers of Synthroid that the medicine has a "history of problems" and cannot be recognized as "safe and effective."[115]

This problem is not unique to Dr. Dong's study of Synthroid. In 1999, the American Association for the Advancement of Science held a wide-ranging conference entitled "Secrecy in Science" at MIT. A number of researchers described similar experiences. In several cases, people who were experiencing detrimental effects to their health at the time would have benefited from the withheld knowledge.[116]

The need for a more critical perspective from the universities is more pressing today than ever before now that many federal agencies are abdicating, at least partially, their responsibility to guarantee the efficacy and even safety of the products and industries that they are supposed to regulate.

Like the Patent and Trademark Office, the FDA has had to rely more on user fees and has taken to pressuring its employees to give faster approval for medicines.[117] The agency proclaims that it wants to be a partner with the pharmaceutical industry. Not everyone benefits from this partnership. Since 1993 the government has had to withdraw seven nonessential drugs from the market because they were suspected as a cause in more than a thousand deaths. Because this number reflects only those cases that have been voluntarily reported, the actual number of fatalities may have been higher. None of these seven drugs were needed to save lives: "One was for heartburn. Another was a diet pill. A third was a painkiller. All told, six of the medicines were never proved to offer lifesaving benefits, and the seventh, an antibiotic, was ultimately judged unnecessary because other, safer antibiotics were available."[118]

When the agency does attempt to pull products from the market, the industry can mount an extensive campaign, often using university researchers, to allow continued sales.[119] Consider the case of GlaxoSmithKline's controversial bowel drug, Lotronex. After five patients taking the drug died, the company voluntarily withdrew it. Shortly thereafter, the company informally approached the FDA to renew the agency's approval of the drug, a strategy that suggests that the company's main concern was financial liability rather than the welfare of patients.

Janet Woodcock, director of the agency's Center for Drug Evaluation and Research, seemed more than willing to comply with the company's objective. She wrote that the "FDA is committed to working with pharmaceutical sponsors to facilitate the development and availability of treatment options for patients with IBS [irritable bowel syndrome]." The letter did not contain a single word of sorrow or regret for the families of those who had died.

An editorial in the British medical journal *The Lancet* reported what followed:

> FDA officials took an increasingly hard line towards their own scientists. Yet new data acquired since the November withdrawal only strengthen the view that Lotronex should not be made widely available again. A further internal review of the incidence of ischaemic colitis among women taking Lotronex suggests that the company may have seriously underestimated the hazards of the drug. And additional adverse reports obtained by Public Citizen show rising numbers of cases of ischaemic colitis and severe constipation in women who continued to take Lotronex.[120]

The editorial concluded: "This story reveals not only dangerous failings in a single drug's approval and review process but also the extent to which the FDA . . . has become the servant of industry."[121] An independent university research system might offer a counterweight to the regulatory agencies that now serve the corporate sector, but, alas, the universities are now beholden to the same corporations.

With critical academic voices silenced, who will protect the public from even worse dangers?

Self-Interest vs. Science

Besides postponing or even forbidding formal publication, entrepreneurial considerations encourage researchers to withhold information from others, even when they are requested to share information. One survey of a stratified, random sample of 2,366 faculty in 117 U.S. medical schools conducted in 1996-97 found that 12.5 percent of researchers in medical schools had been denied access to other academic investigators' data within the previous three years. Researchers who were most likely to be victims of data withholding were those who had withheld research results from others, published more than 20 articles in the last three years, applied for a patent, or spent more than 40 hours per week in research activities.[122]

In addition, as the academic environment becomes more entrepreneurial, disinterested scientific research must surely suffer. A recent editorial in *Nature* entitled "Is the University-Industrial Complex Out of Control?" suggests the scope of this problem. According to the editorial, "One-third of all the world's biotechnology companies were founded by faculty members of the University of California."[123]

Even when the extracurricular activities of these academics permit them to do research, they commonly study subjects in which they have economic interests in the outcomes. As a result, even without corporate intervention, they have good reason to distort their findings. One important study constructed a database of every article published in 1992 by 14 leading life science and biomedical journals that had a first or last author who was affiliated with an institution located in Massachusetts. This study attempted to determine the frequency with which the authors and articles satisfied the condition of possessing a financial interest. Then it examined the articles for any disclosure of financial interest. An author was said to possess financial interest if he or she was a member of a scientific advisory board of a company that developed products related to the scientist's expertise, was listed as the inventor on a patent or patent application for a product or process closely related to the scientist's publication under review, or served as an officer, director, or major shareholder of a for-profit corporation involved in commercial activities related to the scientist's field of expertise.[124]

The study included 1,105 journal authors. Of these, 112, or 10.1 percent, were listed as inventors on patents or patent applications that correlated to a published article in the study. In addition, 69 authors, representing 6.2 percent of the total, were scientific advisory board members of Massachusetts biotechnology companies. Fifteen authors, or 1.53 percent of the author population, who served as company officers, directors, or were a major shareholder, had at least one financial interest in their published article.

The study also looked at the distribution of articles. Of the 789 articles studied, 20 percent of the articles had a lead author on the scientific advisory board of a Massachusetts biotechnology company, 7 percent had a lead author who served as an officer or major shareholder of a biotechnology company, and 22 percent had a lead author who was listed as an inventor in a patent or patent application closely correlated with the publication. All in all, 34 percent of the articles had at least one author who met one of the three criteria of financial interest.

These data underestimate the extent of financial interest for several reasons. First, these indicators do not exhaust the possibilities of financial interest. Evidence of personal or familial investment holdings, consultantships, or honoraria were unavailable. In addition, the study searched for scientific advisory boards only for Massachusetts companies. Many scientists in Massachusetts serve on scientific advisory boards of out-of-state U.S. companies or even international companies. Finally, the frequency of financial interest in research has no doubt been increasing since the time of this study.

Another study reviewed literature concerning the safety of calcium-channel antagonists. The study concluded that 96 percent of the authors who supported the use of these drugs had financial relationships with manufacturers of calcium-channel antagonists, while 60 percent of those who were neutral and 37 percent of those who were critical had no such relationship. In addition, all of the supportive authors had some financial relationship with a pharmaceutical manufacturer.[125] In the summer of 1996, four researchers working on a study of calcium-channel blockers—frequently prescribed for high blood pressure—quit in protest after their sponsor, Sandoz, removed passages from a draft manuscript highlighting the drugs' potential dangers, which included stroke and heart failure. The researchers aired their concerns in a letter to the *Journal of the American Medical Association:* "We believed that the sponsor . . . was attempting to wield undue influence on the nature of the final paper. This effort was so oppressive that we felt it inhibited academic freedom."[126]

In another case, Dr. Martin Keller, chair of the psychiatry department at Brown University, earned more than $500,000 in consulting fees in 1998, most of it from pharmaceutical companies, the effectiveness of whose antidepressants he touted in medical journals and at conferences.[127]

Within this entrepreneurial environment, fidelity to the scientific value of the disinterested search for truth quickly recedes. To keep up with the pack, researchers need to show results. Not surprisingly, with the rise of corporate influence, instances of scientific dishonesty in which researchers resort to fraud have become almost commonplace.

This combination of financial interest and corporate control contaminates the entire scientific process. Science is, above all, a way of gathering and organizing information. The erosion of trust in scientific information created by self-interested behavior spreads confusion. Anything that spreads confusion and doubt reduces the value of scientific information. In this sense, science does not merely depend upon communication per se; but to work right, science requires that researchers be confident that the information that they get is accurate, or at a minimum that those who contribute to the spread of scientific information do so honestly.

Once scientists come to mistrust the motives behind the information that other scientists present, they cannot be sure whether any reported result is the outcome of an unbiased search for the truth or was merely contrived to give the preferred answer. The suspicion of misinformation causes scientists to dissipate their energies in needlessly verifying information that they might otherwise reasonably take at face value.

CHAPTER FOUR

CORPORATE SCIENCE AND INTELLECTUAL PROPERTY

Corporate Science: Dulling the Cutting Edge of Discovery

Supposedly, intellectual property rights are necessary to give corporations the incentive to carry out the research and development that is so vital to future technological progress. Are corporations really the proper vehicle to carry out that mission? I will make the case in this chapter that corporate science is antithetical to scientific progress.

Even without intellectual property rights, corporations will continue to do some research as a means to learn about ways of improving what they do. Most of this research will be of little value to anybody outside of the corporation. Of course, companies might occasionally discover something of scientific interest in the course of attempting to improve their operations, but mostly such research will be specific to the needs of the individual corporation. Such research will continue, even in the absence of any intellectual property protection.

For the most part, however, the kind of science that is truly vital to technological progress is more basic science. More often than not, scientists engaged in basic science have no idea what technologies will evolve out of their studies.

I am not discounting the efforts to develop practical inventions, as opposed to basic science. Already in 1835, Alexis de Tocqueville noticed a marked tendency in the United States to ignore pure science:

> Most of the people in these nations are extremely eager in the pursuit of immediate material pleasures. . . . For people in this frame of mind every new

> way of getting wealth more quickly, every machine which lessens work, every means of diminishing the costs of production, every invention which makes pleasures easier or greater, seems the most magnificent accomplishment of the human mind. . . . In America the purely practical side of science is cultivated admirably, and trouble is taken about the theoretical side immediately necessary to application. On this side the Americans always display a clear, free, original, and creative turn of mind. But hardly anyone in the United States devotes himself to the essentially theoretical and abstract side of human knowledge. In this the Americans carry to excess a trend which can, I think, be noticed, though in a less degree, among all democratic nations.[1]

Later, Thomas Edison boldly expressed this attitude, claiming that the inventor will continue to invent "as along as he commands a dollar. This is a natural peculiarity of the inventive mind."[2] This sort of applied science is useful. It can produce useful devices or make a company more productive, but it does not produce the types of breakthroughs that result in revolutionary technological change.

Such basic science is an arena in which playful labor, rather than the narrowly focused work of an Edison, is most effective.[3] In this vein, Alfred Marshall, probably the world's most influential economist of the late nineteenth and early twentieth centuries, observed:

> The chemist or the physicist may happen to make money by his inventions, but that is seldom the chief motive of his work. He wants to earn somehow the means of a cultured life for himself and his family; but, that being once provided, he spends himself in seeking knowledge partly for its own sake, partly for the good it may do to others, and last, and often not least, for the honour it may do himself. His discoveries become collective property as soon as they are made, and altogether he would not be a very bad citizen of Utopia just as he is.[4]

After all, what draws people to science in the first place? I doubt that few, if any, scientists have followed the lure of monetary gain in choosing a scientific career. Most accounts that I have read describe future scientists as children who were captivated by a mystery that they wanted to understand at an early age. As children, these future scientists seem to have recognized what administrators today do not—that scientific exploration can and should be a joyful experience; that the thrill of discovery is an essential element of the scientific process. In the words of David Hull, a philosopher of science, science is "play behavior carried to adulthood."[5] Maybe this situation will change in the future, now that some superstar scientists are beginning to command substantial salaries or to earn fortunes by exploiting the monetary value of their discoveries. Insofar as science is concerned, however, I cannot believe that pure greed will ever be nearly as effective in promoting good science as the playful but passionate labor of serious scientists.

Corporations have little interest in providing the kind of environment in which science flourishes. Most corporations see no reason for encouraging scientific activity that does not contribute directly to profits. For example, many scientists desire to have the opportunity to communicate with their peers by publishing their results in professional journals. Corporations, however, frequently cannot appreciate the importance of publication. One biotechnology executive expressed this perspective, commenting, "why should I let my people publish? It's just a waste of time that could be spent in the search for new drugs."[6]

One survey of postdoctoral biologists in the biotechnology industry found that even with the enormous opportunities for great wealth in this field, scientists willingly make financial sacrifices to continue to participate in the larger scientific processes. Firms that allow their employees to publish in scientific journals pay an average wage of 25 percent below firms that prohibit such activities.[7]

More generally, the rules of corporate science typically do not allow the freewheeling ambling-about that is the hallmark of traditional science. In the words of Joseph Schumpeter, probably the most influential twentieth-century economist on the subject of the economic role of new technology, "Technological progress is increasingly becoming the business of teams of trained specialists who turn out what is required and make it work in predictable ways."[8] In this vein, corporations demand quick results on tightly focused objectives.

In short, the rules of corporate science are antithetical to scientific progress. Yes, corporations will finance some valuable discoveries, but the extent and importance of these discoveries is far less than what the same research effort would have produced without the tight structuring imposed by the short-term profit motive.

In fact, the atmosphere surrounding corporate bureaucracy is one of the least congenial environments for fostering creativity. Consider the attitude of Daniel P. Barnard, research coordinator of Standard Oil of Indiana at the time, described the policy of his corporate employer: "We employ many people who, if left to their own devices, might not be research-minded. In other words, we hire people to be curious as a group. . . . We are undertaking to *create* research capability by the sheer pressure of money."[9]

Some of the most heroic lore of recent science concerns the military achievements during World War II, when scientists created the atomic bomb and constructed some of the first computers. Could these accomplishments represent the benefits of organized and structured research? Alternative explanations seem to be more reasonable. First, many of these scientists saw themselves as providing a strategic contribution to a war in which they fervently believed. At the same time, they were able to work at the forefront of science. Corporate-style management would have been superfluous in prodding them to higher achievements. The military goals might have focused their attention, but the intensity of their motivation was far more powerful than we could expect from a salaried employee working on a project defined by a corporate bureaucracy.

Alternatively, the accomplishments of those wartime scientists might merely reflect the fact that the government had gathered together many of the most powerful minds of the time. Enormous scientific progress was inevitable once such people, who lived and breathed science, had the opportunity to spend so much time together supported by virtually unlimited research budgets. Again, corporate-style management would have been superfluous, if not constrictive.

Patent Races and Reverse Engineering

The bizarre applications of intellectual property rights that I discussed earlier are symptomatic of a future in which corporations will dissipate more and more of society's resources in wasteful litigation over intellectual property rights. Intellectual property rights are counterproductive in another respect: Because of their enormous monetary value, corporations throw themselves into "patent races," in which a number of firms try to be first to win the exclusive right to a widely understood objective.

The law encourages these patent races. Recall how the inability to determine who deserves how much credit for a particular invention leads to the adoption of a short cut in which the first-to-file wins the patent regardless of the contribution of others. This first-to-file rule encourages the waste of money, time, and energy.

To begin with, firms duplicate each other's efforts. In addition, pressuring research to be done quickly is costly, both in economic and personal terms. Tracy Kidder's *The Soul of a New Machine* may provide the best case study of the personal side. This book tells of the toll taken on a group of young engineers racing to bring a new computer to market.[10]

Frequently, the purpose behind the development of a patent is not to protect a firm's hard-won scientific and technical achievements. Instead, the firm knows that other companies stand ready to sue it for violation of their intellectual property rights. By acquiring as many patents as possible, the company has a better chance of countersuing to neutralize such suits. This practice is common among producers of computer chips.

Bronwyn Hall and Rose Marie Ham of the University of California, Berkeley, describe how corporate research directors in the semiconductor industry set quotas for the number of patents that they expect to win each year.[11] No matter that these patents may be for methods or procedures that have been in place for some time within the firm; their purpose is to create a thicket of legal barriers for companies that might be inclined to challenge some of the firm's intellectual property rights.

In effect, then, these patents do nothing to stimulate research and development. In fact, efforts to secure patent rights will certainly distract researchers from doing useful scientific work. Instead, these patents are nothing more than bargaining chips in an effort to forestall potential litigation against the firm or to reinforce litigation that the firm might want to pursue.

Once a particular firm has won a patent race, a new form of duplicative waste commences. Other companies will attempt to reverse-engineer the patent. In effect, they will attempt to copy the achievements of the first firm while avoiding the charge of violating that company's intellectual property rights.

Earlier, I discussed surveys of corporate executives and research managers regarding the ease with which a company can reverse-engineer products that other firms have patented. Theoretically, the company cannot merely copy the product. It has to follow a procedure that indicates that it has done the work on its own.

For instance, when one company creates a new computer chip, engineers working for a competitor will examine the chip. They supposedly pass their information on to another group of engineers without telling them exactly how the chip works, only what kind of operations it performs. This second group of engineers is expected to replicate those operations without any prior knowledge of how the first company initially achieved its task. Since a so-called Chinese wall prevents anything but information about the ultimate functions of the chip to move between the two groups of engineers, the firm can claim to have created the new chip without violating the intellectual property rights of the other firm. Although the engineers who copy the chip pretend to have known only the functions of the chip, not the way the first company had originally engineered these functions, the first company will challenge this claim, accusing the engineers of directly copying the chip.

As a result, besides the wasteful duplication of efforts, reverse engineering leads to endless litigation concerning the existence and effectiveness of the Chinese wall. The copier must be able to show a detailed paper trail of flow charts, time sheets, and results of computer simulations.[12] This procedure is not unique to the semiconductor industry.

If the same scientific effort were carried out cooperatively, the people involved could have achieved the same result with a fraction of the resources. But then, cooperation is antithetical to intellectual property rights, even though it may be a natural behavior for people sharing an interest in developing technology.

For example, Eric von Hippel, a professor of management at MIT, insisted that engineers in many old-line industries had surreptitiously shared information in the past.[13] This cooperative relationship promoted technological progress, although the industries that he studied were admittedly not ones in which intellectual property rights played an important role. Nor were they industries on the cutting edge of technology.

Workers as Intellectual Property

One of the main contradictions in the case of the proponents of intellectual property concerns the weird legal standing of research workers. Recall that the initial justification of patents was based on the encouragement that they gave to individual inventors. Today, of course, the overwhelming majority of patents issued go to corporations rather than individuals.[14] This transformation obliterates the initial rationale of patent rights.

While a corporation supposedly requires the promise of potential intellectual property rights to make it commit to performing research, the actual researchers themselves cannot look forward to becoming owners of the intellectual property that they create. In effect, the corporations treat their organizations as a commons to which they expect their employees to assign the rights to all of the intellectual property that they create.

This transformation of the individuals also appears in economic theory. According to the perspective of conventional economics, individual researchers accumulate human capital through education or experience. This human capital supposedly represents a factor of production, comparable to physical capital, such as a machine that amplifies the contribution of the worker. Ownership of human capital allows workers to earn a higher wage, but they have no more right to the product that they produce than the machine does.

Now imagine that society were to attempt to apply the same logic to the intellectual property of the corporation that the corporations use to justify ownership of the intellectual property of their workers. Suppose that society offered to hire firms as researchers the same way that firms hire workers. Under this arrangement, firms would have the same rights that employees now have within firms. Consequently, firms would sell the human capital accumulated within the company for a fixed price to society. Would corporations willingly accept this new system, under which society as a whole would have control of the fruits of their research? The corporations would undoubtedly respond that such incentives would be insufficiently sharp to induce them to make such wonderful contributions as they now do.

Why then do the corporations not let their workers have the same sort of incentives that they demand for themselves? They would respond, as they already have in the courts, that the problem of assigning property rights to individual workers would create intolerable confusion.

The corporations would be correct in arguing that they would have no way of determining how much each individual worker contributed to a particular process since scientific advance generally is a result of teamwork. Of course, the same argument holds for science in general. Science is an interactive process. Scientists do not work in a vacuum. Nobody, absolutely nobody, can tell how much each participant contributed to the eventual outcome of any scientific study.

Just as the corporations rightfully worry that time and resources would be wasted in attempting to assign specific values to the effort of each individual worker, society should be concerned about the enormous costs arising out of disputes among corporations over intellectual property. Just as the corporation represents the collective repository for the information generated within the organization, one can make the case that society as a whole should be the collective repository of the scientific efforts of the community as a whole.

The patent system circumvents the problem of determining relative contributions by assigning ownership to the first one—either a corporation or an individual—that files a patent, regardless of the extent of the contribution of that agent in developing the science behind the patent. This approach leads to the wasteful patent races that I just discussed.

The Challenge of Corporate Ownership

One might argue that corporations provide the framework in which individuals make contributions to the collective intellectual property that the corporations own. In fact, even when a corporation makes no contribution whatsoever, it still is more likely than not to claim ownership of the intellectual property that an individual employee contributes.

For example, commercial publications often purchase stories from freelance writers. The publication contributes nothing to the creation of the story. It merely purchases the story as a commodity from writers who are not employees. After it publishes the material, it may resell it to third parties who distribute it electronically. The freelancers sell their work for one purpose—the initial publication—only to see it resold for another purpose without their permission.

In a legal test of this arrangement, Jonathan Tasini, the president of the National Writers' Union and the lead plaintiff, sued the New York Times Company, Newsday, and Time Inc. and their co-defendants, University Microfilms International and Mead Data Central Corporation, the former owner of the Lexis-Nexis databases. The plaintiffs lost in a federal district court, but on September 24, 1999, the Second Circuit Court of Appeals reversed the decision, ruling that the reuse of freelance work on databases and CD-ROMs without the authors' express permission constitutes copyright infringement.[15]

Publishers have responded to the suit by requiring, as a condition of the purchase of a story, that the freelancer assign to the publisher the electronic rights to that story. In other words, even though the company contributes nothing to the development of the story, they still maintain the right to claim it as intellectual property. The freelancers have taken at least one newspaper, the *Boston Globe,* to court, challenging this practice.[16] The *New York Times* responded by issuing an internal memo recommending that the paper not do further business with the 11 plaintiffs and 2 other writers who filed a different suit.[17]

Well, you might think, the buyer has a legitimate right to determine the conditions of the contract. Before you take that argument too seriously, take a look at the conditions to which you agree when you purchase a piece of software. Do you seriously think that you could go to Microsoft and demand certain rights as a condition of purchasing their product?

Similarly, when Monsanto sells seeds to a farmer, the corporation makes the farmer agree to a stringent set of conditions about how the seed will be used. In fact, the company even denies that it has sold the seeds to the farmer, claiming that it has merely rented the rights to its intellectual property. If the company suspects that the farmer has violated the conditions of the "purchase," it brings the legal system down heavily on the farmer.

Monsanto recently won a case against a Canadian farmer who insisted that the genetically engineered characteristics of the plants that he grew were the result of pollen that had drifted from neighbors' fields. The case was even more curious because Monsanto engineered its seeds to have one particular characteristic: the plants were resistant to Roundup, Monsanto's best-selling herbicide, so that farmers could spray large quantities of herbicide without damaging the crop. This particular farmer, Percy Schmeiser, did not farm in such a way that this herbicide resistance would be of any use to him.

The decision ruled that "the source . . . is really not significant." The judge also decided:

> In my opinion, whether or not that crop was sprayed with Roundup during its growing period is not important. Growth of the seed, reproducing the patented gene and cell, and sale of the harvested crop constitutes taking the essence of the plaintiffs' invention, using it, without permission. In so doing the defendants infringed upon the patent interests of the plaintiffs.[18]

The farmer had the responsibility to call Monsanto and tell them that they should reclaim their intellectual property by removing the plants from his fields.[19] Roger Hughes, the Monsanto attorney, even argued, "Whether Mr. Schmeiser knew of the matter or not matters not at all."[20] The judge did not say whether he was willing to go that far.

Monsanto had 475 seed piracy cases nationwide as of 1998.[21] After its heavy-handed tactics came under criticism, it shrouded its legal activity in secrecy. I telephoned Monsanto's public relations office, but the representative refused to release any information to me. None of the people who follow the company closely have had any success in finding recent information about the extent of this activity.

What is common in all these cases is that the control of the intellectual property goes to the most powerful party, whether it be the purchaser of a story or the seller of software or seeds.

Intellectual Property and the Distortion of Technology

Admittedly, intellectual property rights can create an incentive for the creation of intellectual property, other things being equal. The problem is that other things are not equal.

Even when intellectual property rights were far weaker than anything that we know today, they still might have impeded scientific progress. For example, most economists today regard the creation of the railroad system as the most revolutionary industrial development of the last half of the nineteenth century. The railroad literally transformed the face of the world, at the same time creating enormous demand for great industries such as oil and steel.

The highly regarded economic historian Joel Mokyr observed that patents raised difficulties in the creation of the basic technologies that made the railroads possible. For example, James Watt held a patent that covered noncondensing engines that "wrought by the force of steam only." This claim effectively blocked the development of a high-pressure engine, even though Watt himself firmly opposed such engines and did not plan to develop them. Another engineer held a patent that prevented Watt from converting the reciprocal motion of his steam engine into circular motion, although he eventually circumvented the obstruction.[22]

Frederick Scherer, the preeminent authority on the economics of industrial organization, claims that James Watt's refusal to grant licenses for his steam engine patent impeded the work of Jonathan Hornblower, Richard Trevithick, and others working on the development of high-pressure engines. Instead, they had to wait until Watt's patent expired in 1800. Access to this patent was essential in developing the basic technology for steam-powered rairoads and boats. Scherer claims that this obstacle may have had some effect in delaying the introduction of steam locomotives and steamboats.[23] So, one patent impeded Watt, while his own patent created a barrier for others.

Scherer might be wrong. Watt's patent may not have obstructed technology at all. Indeed, Scherer does not push his point very hard. Nor did Mokyr. Obviously, absolute proof in such speculation is virtually impossible. Even so, Scherer's hypothesis suggests that the barriers imposed by intellectual property rights may significantly distort economic progress. That he could even propose such a hypothesis for a time when intellectual property rights were so much weaker than they are today is a telling point.

Also, as I mentioned earlier, the negative consequences of intellectual property rights accumulate as the networks of property rights become more complex. In the earlier stages of development, the positive incentives to invention probably outweighed the negative consequences. Early societies lacked the alternative sources of research offered by the contemporary scientific infrastructure, consisting of a complex array of advanced university research labs and programs for training graduate students. In addition, at the time, the great innovations depended more on invention than on basic scientific research.

Today, intellectual property rights are much stronger and dependence on science far more crucial. The loss of resources into the sinkhole of wasteful litigation, along with the diversion of scientific energies into reverse engineering and patent races, impose an appalling cost on society. Other costs, although more difficult to measure, are probably even more destructive than litigation.

Perhaps the most important of these costs is the spread of secrecy within the scientific process. Such costs are incalculable, as is the transformation of the university from a place that nurtured pure science into an entrepreneurial source of income from more applied technologies.

Intellectual property reduces economic potential in other ways. To begin with, companies go to great lengths to develop or modify technologies in ways that have no purpose except to protect intellectual property. For example, in the 1970s and early 1980s, many software companies went to great lengths to create measures to complicate efforts to duplicate software in an effort to protect their intellectual property. In the process, they made their own software more inconvenient to install, if not to use. Even today, the installation of software often requires the keying in of long sequences of numbers, creating a nuisance every time that software needs to be reinstalled.

In other cases, corporations spend money to make their products less useful to capture more rents from their products. For example, Intel purposely disabled the math coprocessor in the lower cost Pentium SX chip to encourage customers to purchase its more expensive products. The Recording Industry Association of America (RIAA), composed of member companies responsible for almost 90 percent of the prerecorded music produced in the United States, goes to great lengths to make the duplication of music more difficult.

For example, in the 1980s the RIAA pressured Congress to limit the capacity of digital audio technologies. Congress dutifully responded with the Audio Home Recording Act of 1992, which required producers of digital recording devices to install a chip in their system that implements a code-based system to monitor the copies of any copy made on each machine.[24] More recently, Creative Labs developed a new portable MP3 player with a built-in FM radio that can record radio music. The company had to engineer it to record with low quality to avoid being sued by the recording industry.[25] In sum, industries that produce technologies capable of reproducing recorded music must include technologies that disable the capability of duplication, or at least impose difficulties on those who would like to do so. As a result, consumers must pay the extra costs of developing and producing technologies designed to restrict products' usefulness.

IBM, Intel, Toshiba, and Matsushita Electric have jointly developed a new technology for the convenience of copyright owners. The technology, known as Content Protection for Recordable Media, or CPRM, allows content producers to specify how many times a consumer can copy a given file. When you buy and download, your MP3 player would use the rights-protection system and the serial number already on your memory card or disk to encrypt the file and create a unique "key" for it. That key lets the music player know whether or not the file is stored on an authorized disk or memory chip. When you want to listen to your album, the player checks for the digital key; if everything matches up, the file is decrypted and your music will begin to play. The copy-protection system will not work unless it is deployed in the original files, in storage media, and in media players. To add insult to injury, consumers will have to pay for the necessary hardware, as well as royalties to the owners of the technology to interfere with consumer behavior.

Microsoft has built its new XP operating system in a way that severely limits the quality of sound that can be recorded as an MP3 file. The system is designed to work with non-MP3 music-software formats that include technology known as digital-rights management, which can "lock" copyright-protected songs and make it difficult for consumers to share music. Dave Fester, a general manager in Microsoft's Digital Media Division, arrogantly asserted, "We think at the end of the day, consumers don't really care what format they [record] in."[26]

New DVD players provide a Regional Code Enhancement, but this enhancement does nothing for you. Instead, it prevents you from watching a DVD sold in another region. For example, region one discs only play in the United States.

What pressing reason demands Regional Code Enhancement? Movie studios maximize their profits by staggering the release of titles on videotape and DVD in different regions following the schedule of their cinema releases. So they want to prevent people from watching a film before it is released in their region.[27] University of California, Berkeley, Law Professor Pamela Samuelson observed, "These groups seem to believe they are so important to America that they should be allowed to control every facet of what Americans do with digital information. They also seem to think they are entitled to control the design and manufacture of all information technologies that can process digital information."[28]

The efforts to cripple the usefulness of technology do not always succeed. For example, Hollywood faced a crisis in new technology. The president of the Motion Picture Association of America, Jack Valenti, in testifying before the House Judiciary Committee, roared, "The growing and dangerous intrusion of this new technology [threatens an entire industry's] economic vitality and future security. [The new technology] is to the American film producer and the American public as the Boston Strangler is to the woman alone."[29] The alert reader might notice the dated reference to the Boston Strangler. Mr. Valenti was testifying in 1982 against the threat posed by videocassette recorders.

You may have noticed that the industry is still earning healthy profits, despite the introduction of this murderous technology. Was Mr. Valenti correct or was the industry saved by the inclusion of technology to making the copying of videos more difficult? In the current regime of intellectual property rights, Mr. Valenti would certainly be more likely to succeed in impeding the introduction of the VCR.

At the time of this writing, the industry is pressing the Federal Communications Commissions to require virtually every digital device to include special circuitry to make copying future digital television signals all but impossible. The industry plan would render the existing generation of already very expensive high-definition television sets incapable of viewing many digital programs. It would also prevent households from recording shows on their VCRs. The construction of all of these technological impediments to using technology goes hand in hand with the legal actions to crush those who would provide more convenient ways to copy music, as in the case of Napster.

Restrictions on intellectual property do not make sense, even by the narrow criterion of profits. Napster actually seemed to increase sales of CDs, by allowing people to become familiar with new music. Video rental income now rivals box-office receipts for movies.[30] In the case of software, copying often helps to improve the product and, more important, to establish a program as a standard.[31]

I would like to mention a final distortion that the current state of intellectual property law creates. The present system imposes a perverse incentive for pharmaceutical companies. In the words of a United Nations report: "Vaccines are the most cost-effective technologies known in health care, preventing illness in an one-time dose. But they generate smaller profits and have higher potential liabilities than treatments used repeatedly."[32]

According to the logic of intellectual property, testing or lifelong treatments promise far more profits than a permanent cure. Such concerns would not occur to a disinterested scientist, but they can be of the utmost importance to a profit-minded corporation. As a result, pharmaceutical companies prefer to invest in medicines that require continual application for the rest of patients' lives or procedures for repeated testing rather than searching for a cure for a disease.

The Terminator Gene as a Protector of Intellectual Property

Perhaps the most egregious example of intellectual property rights leading to the introduction of an innovation that makes a technology less productive is the so-called terminator gene. Delta and Pine Land Co., a firm that Monsanto was in the process of acquiring, together with the U.S. Department of Agriculture (USDA), developed this gene to prevent farmers from saving seeds from their harvest to plant the next year.

Traditionally, farmers kept seeds from their previous harvest. They would replant these seeds and even trade them with other farmers. As a result, seed companies could make a one-time sale, but they would have difficulty in making farmers purchase new seeds each year. So, unlike the typical technology company, which uses patents to prevent its competitors from duplicating its product, the seed companies fear competition from their own customers.

A crop with the terminator gene will grow and flower, but its seeds will be incapable of reproducing. The value of this invention to the seed companies is obvious. Melvin Oliver, the primary "inventor" of the terminator technique, said, "the need was there to come up with a system that allowed you to self-police your technology, rather than trying to put laws and legal barriers to farmers saving seeds, and to try to stop foreign interests from stealing the technology."[33]

The terminator seed has given rise to protests around the world. This response is not surprising since the technology will impose serious costs on poor farmers around the world, in many cases threatening their very livelihood. Traditionally, peasant farmers throughout the world have managed to scrape by through finding ways to minimize their dependence on the market. The emergence of this application of genetic engineering threatens to put an end to these practices.

The terminator gene has an interesting precedent. Although the seeds of hybrid corn plants can reproduce, they produce a very low yield. This characteristic offers a significant advantage to breeders. Indeed, Edward East and George Shull, who originally developed the hybrid technique, wrote that the inability to replant hybrid corn profitably was one of its most attractive features: "It . . . is the first time in agricultural history that a seedsman is enabled to gain the full benefit from a desirable origination of his own or something that he has purchased."[34]

Like the terminator gene, the commercial production of hybrid corn seed required a massive government effort. Then, Henry Wallace, the son of the United States Secretary of Agriculture, took a small amount of the seed lines, grew them on a commercial scale, and went into the seed business, eventually becoming the largest seed producer in the United States.[35] In effect, the creation of hybrid corn transformed seed from being the collective product of many thousands of farmers over hundreds of generations into a commodity that could be sold by an individual company.

The terminator gene differs from hybrid corn in one important respect. The quantity of hybrid seeds presently used by the poorest farmers around the world is negligible. The terminator gene, in contrast, threatens to raise the commodification of the agricultural heritage to an undreamt-of level.

The spread of the terminator gene will threaten small farmers in a number of ways. First, they will have to return to the market each year. In addition, they will probably have to rely on credit to pay for their seeds. The terminator gene also poses a risk for the world food supply. The traditional agricultural practice of saving seeds helped to maintain a genetic variability that benefits farmers as well as the world as a whole. The widespread use of the purchased seeds threatens to create excessive genetic homogeneity. Without genetic variability, when a serious outbreak of a plant disease occurs—and eventually these outbreaks are inevitable—the crop of the entire region is put at risk. Poor farmers, without access to funds to tide them over, are most likely to fail during such crises. Moreover, as the natural genetic diversity disappears, the world will become dependent on the seed companies to create diversity through the genetic modification of seeds.

True, plant breeders might also turn to a handful of seed banks in which a number of varieties are stored. Unfortunately, these seed banks contain a tiny fraction of the total variety in use today. Moreover, since the seed banks rely on electrical cooling systems, concern exists about the present state of the seed banks and their vulnerability to an interruption in their power supply.

Why, then, would any farmer adopt such a destructive technology? Public or commercial lending agencies often pressure farmers to adopt "modern" farming practices as a condition of obtaining necessary credit. In addition, educational institutions persuade farmers that "modernization" is the road to prosperity.

Perhaps most troubling is the support of the U.S. Department of Agriculture for this research. Why should scarce public funds be used to subsidize the intellectual property rights of a tiny number of seed companies at the expense of poor farmers? Not surprisingly, the government's response is that the protection of intellectual rights will stimulate future advances in the genetic modification of crops.

Of course, doubts persist about the wisdom of genetically modified crops. Already, some modified genes have been able to jump from one species to another. For example, a three-year study by Professor Hans-Heinrich Kaatz at the University of Jena found that the gene used to modify rapeseed oil had transferred to bacteria living inside honeybees.[36]

The approach of the USDA is especially disturbing. This agency once ruled over what perhaps was the most extensive program of public science in the history of the United States. The history of the privatization of hybrid corn suggests that this program was not above reproach. The USDA never emphasized basic science; it has long been heavily weighted toward the needs of the largest commercial farmers, especially the corporate farmers.

Nonetheless, its network of farmers and scientists had a tradition of sharing seeds to promote the improvement of crops. Today, researchers have difficulty obtaining seeds from each other. A researcher at the University of Costa Rica developed a strain of rice that is resistant to a virus that creates serious problems in the tropics, but before the university can release the plant to the public, it must first obtain clearance from holders of as many as 34 patents.[37]

The terminator gene probably represents the natural culmination of a growing emphasis on genetic engineering. The main advantage of this technology is not so much a promising array of plant improvements. Instead, genetic engineering improves the chances of capturing lucrative intellectual property rights.[38]

Intellectual Property in the Pharmaceutical Industry

The pharmaceutical industry is one of the main beneficiaries of intellectual property rights. Certainly, the pharmaceutical industry would have a hard time surviving in a free market. It depends upon substantial research costs to develop a product that costs virtually nothing to produce. Because of these characteristics, free and open competition would inexorably drive most pharmaceutical companies into bankruptcy.[39] The present system, however, has been overly generous in granting special privileges to this industry.

The prices of medicine have long been a source of displeasure. Adam Smith himself wrote: "Apothecaries' profit is become a bye-word, denoting something uncommonly extravagant."[40] Smith rose to the defense of the industry, justifying its profits as legitimate compensation for the druggist's skill.

Today, even principled defenders of intellectual property rights might be taken aback by the exorbitant rates of profit that this industry collects.[41] Those who closely follow the pharmaceutical industry often refer to it as the most profitable legal business on the face of the earth. For example, the pharmaceuticals top the list of all industries in the Fortune 500. The industry boasted a reported profit on sales of 18.5 percent in 1998. Amgen led the industry, reporting 1998 profits as 33 percent of sales. Among the 11 other pharmaceutical companies that made the Fortune 500 list, 4 made profits in excess of 20 percent of sales, and another 3 earned in excess of 15 percent.[42] With the population aging, the industry has reason to expect even greater profits.

The pharmaceutical industry enjoys huge government largess. The government writes tax laws to benefit the industry. For example, the average tax rate of major industries from 1993 to 1996 in the United States was 27.3 percent of revenues. During the same period the pharmaceutical industry was reportedly taxed at a rate of only 16.2 percent.[43] Moreover, government laboratories develop new products, and then turn them over to the pharmaceutical companies, which magically transform public research into private intellectual property. Finally, the government graciously distorts the overly generous intellectual property laws to give special consideration to the pharmaceutical companies.

The industry contends that these high costs are necessary if it is going to continue to develop new lifesaving products. A widely cited study estimates the average cost of bringing a drug to market at $231 million.[44] The authors of the study had close ties to the industry, which is reluctant to release its data to other sources. The industry claims that the number should now be more than doubled to $500 million because of inflation and the extra testing required by the Food and Drug Administration (FDA).[45]

The federal Office of Technology Assessment regarded the $231 million estimate as an upper bound.[46] Only half of the $231 million represents funds actually spent for development; the other half is what economists call opportunity cost—the return that the companies could have expected if they had invested the money instead of tying it up in developing the drug.[47] In addition, the government provides a substantial part of what the report assumes are the costs of developing a drug.

At one point, the industry did provide an accounting of some of its costs of developing drugs. Under the Orphan Drug Act of 1983, the government granted the industry financial support to promote the development of medicines for rare disorders. The government paid half the cost of clinical trials and gave the developer a seven-year exclusive monopoly in marketing the product, whether or not it had a patent or even if it were unpatentable. James Love, director of economic studies at the Center for Study of Responsive Law, analyzed the costs reported by companies that qualified for these subsidies. The average out-of-pocket expense was $3.2 million in 1995 dollars per approved drug, compared with the study's estimate of $24.5 million.[48]

In addition, a recent *New York Times* study discovered that much of the research by drug companies is aimed not at innovation at all, but rather at developing medicines that have the same function as similar drugs made by rivals. It reported that officials at the FDA say that of the 90 new drugs approved in 1992, only about 40 percent were a significant advance over medicines already available. Finally, in 1992, the industry spent $1 billion more on promotion of its drugs than on research and development.[49] Recall the earlier estimate that by 1997, 42 of the top 100 drugs were imitations of existing drugs, an effort comparable to the reverse engineering of computer chips.

Since that time, spending on promotion has soared. Leading the pack is Pfizer, a company that boasts a worldwide army of 20,000 sales representatives. William C. Steere Jr., the company's chairman and chief executive, began his career there as a sales representative. A writer for the *New York Times* described Pfizer as "probably the first in the industry to transform itself so clearly from a research-driven company to one that operates more like Procter & Gamble, the maker of Tide."[50]

Pfizer is not alone in its emphasis on marketing. The *Wall Street Journal* describes the increasing dominance of marketing at Merck:

> For decades, Merck's marketers hadn't been allowed anywhere near scientific planning meetings. [Eventually, the company] started to change this, persuading scientists to accept marketers in their midst, promising that they wouldn't speak. Then speaking became allowed, but not encouraged. [More recently], however, the marketers have become deeply involved in many of the scientists' development decisions, though they still have no involvement in early-stage research issues.[51]

Merck has also "created teams of marketing, manufacturing and research people that now plan far ahead."[52]

AIDS Action reports that in 1998, the 15 largest pharmaceutical companies collectively spent almost three times more on marketing, advertising, and administration ($68 billion) than on research and development ($24 billion). No one who has access to a television set will be surprised that in 1998, direct consumer advertising on drugs increased at a rate three times higher than spending for research and development (54 percent vs. 17 percent).[53]

In 1998, according to the research firm, Competitive Media, Schering-Plough spent $136 million advertising just one medicine, its allergy drug Claritin. The cost of this advertising blitz exceeded what Coca-Cola spent to advertise Coke or Anheuser-Busch spent to advertise Budweiser. The company's advertising was successful. It has made Claritin the dominant allergy drug, with $1.7 billion in sales in the United States, despite the absence of any evidence that Claritin functions better than competing nonsedating antihistamines like Allegra, which has sales less then one-third of Claritin's.[54]

Television advertising is only a part of the story. The pharmaceutical companies spend 60 percent more on marketing directly to doctors than they do on advertising directly to consumers.[55] The pharmaceutical companies command an army of more than 68,000 salespeople, one for every eleven doctors in the United States.[56] To make their products even more attractive they offer doctors cash, as well as everything from free meals, Valentine's Day flowers, books, CDs, manicures, pedicures, car washes, and bottles of wine.[57] One company, Pharmacia, spent 40 percent of its overall revenue on marketing and administrative expenses in 1999, more than twice what it spent on research. In the last year alone, it increased its global sales force by 30 percent, to 6,500 people.[58]

Presently, the pharmaceutical companies, despite their stated commitment to developing new products, appear to have few new products in their pipeline. The number of novel drugs approved by the FDA peaked in 1996 at 53, compared to just 35 in 1999, and 16 through the first half of the year 2000.[59] In order to shore up their position, the pharmaceutical companies are frantically engaged in a series of megamergers.[60] These mergers will, no doubt, strengthen the political clout of the industry, allowing it to extract even more benefits from its intellectual property rights.

Dean Baker of the Center for Economics and Policy Research made a few rough calculations concerning the costs of intellectual property in the pharmaceutical industry. Presently, people in the United States spend close to $100 billion a year on prescription drugs. In the absence of patent protection, Baker estimated that the cost of these drugs would fall to less than $25 billion—a savings of more than $500 a year for every household in the country. By contrast, the proponents of deregulation in the airline, trucking, and telecommunications industries put the gains from each of these policies in the neighborhood of $10 billion to $20 billion annually.[61]

Yes, but what about the great medical advances that arise out of the efforts of these companies? Dean Baker observed:

> According to its own data, the pharmaceutical industry funds only 43 percent of medical research in the United States. The federal government funds close to a third of all medical research, primarily through the National Institutes of Health. Universities, private foundations and charities account the rest. These other methods of funding research have a proven track record. This research has produced a long list of major medical breakthroughs, including the discovery of penicillin, the polio vaccine and AZT (though not its use as an AIDS treatment). In just the past two months, NIH researchers developed a vaccine that will prevent the transmission of AIDS through breast-feeding, and a use for aspirin for people undergoing heart surgery. The industry is presently spending approximately $20 billion a year on research. Some portion of this spending, probably in the neighborhood of one-third, is devoted to researching copycat drugs. But in the absence of the patent the amount of research spending that would have to be picked up in the absence of patent protection comes to approximately $13.3 million a year. This amount is approximately equal to what state and federal governments could expect to save on Medicare and Medicaid payments for prescription drugs in the absence of patent protection. . . . In other words, this would allow the patented price of drugs to fall to a free market price that on average would be less than 25 percent (and in many cases less than 5 percent) of the patent-protected price.[62]

For families that enjoy relatively good health, high drug costs might seem to be a fairly abstract problem; however, for people of modest means with chronic conditions that require continual reliance on expensive drugs, the costs of medicine can represent an enormous burden. Busloads of people routinely cross the border into Canada and Mexico just to buy their medicine at a lower cost. For others, the expense of drugs requires that people must choose between their medicines or their rent or food.

The pharmaceutical companies would have the public believe that exorbitant costs of developing drugs necessitate the high prices. This claim rings hollow indeed. Certainly, the practice of repricing veterinary medicines for human use casts considerable doubt on that attempted justification. For example, back in 1992, Johnson and Johnson was charging $1,250 to $1,500 for a year's supply of the drug Levamisole, used to treat colon cancer. The price was not particularly out of line with other medicines. What made this case unusual was that the drug, which cost about $6 per pill, was a slightly reformulated version of a pill to fight worms in sheep that cost veterinarians only about 6 cents.[63] When questioned about this price gouging by ABC's PrimeTime Live, Robert Gosson, a corporate vice-president for the company, responded, "A sheep farmer probably would not pay $6 a pill," while "someone dying of cancer who pays $1,200 for a treatment regimen, whose life is saved, is getting one of the most cost-effective treatments ever."[64]

This impeccable logic describes a world in which firms charge whatever they can regardless of development costs. Although more is involved than just the pharmaceutical industry, an estimated half a million families filed for bankruptcy during 1999, in part to deal with the financial consequences of an illness or accident.[65]

The comparative performance of the health care system of the United States is a fascinating indicator of the success of the present system of intellectual property in medicine. Although the United States spends a larger share of its gross domestic product on health care than any of the 191 countries that the World Health Organization studied, it ranks just thirty-seventh in the world, slightly below Cuba, which is ranked thirty-fifth.[66]

Stifling Invention in the Pharmaceutical Industry

While the pharmaceutical industry claims that it needs protection of its intellectual property rights to further the development of lifesaving technologies, its record is mixed at best. During the dawn of the biotech era, when the industry was more forthcoming, *Business Week* discussed the dilemma of Genetics Institute, Inc., of Cambridge, Massachusetts. The company had to chose between two versions of a clot-dissolving drug to develop, since it only had money for one. The scientists selected the drug that tested much better. The attorneys pushed for the other drug. Although it did not test as well, it had the broadest patent. The attorneys won.

Bruce M. Elsen, the company's patent counsel, said, "Researchers used to be up in arms if such crass decisions were made. . . . [But now] the strength of the patent position is a leading factor in what research to pursue."[67]

Martha Luehrmann, formerly with the Lawrence Berkeley National Laboratory, circulated an email about an even more chilling example of corporate irresponsibility. Amgen produces and owns most of the rights to a drug called erythropoietin, or EPO, a naturally occurring hormone, which has proven to be extremely effective in encouraging the development of oxygen-carrying red blood cells and has meant life for many anemic people, including premature infants and those with anemia due to kidney failure, other diseases, or surgery.

Parenthetically, EPO is also the performance-enhancing drug of choice of bicycle racers and other long-distance athletes, because the extra red blood cells increase their oxygen carrying capacity. Unfortunately, the additional red blood cells thicken the blood so much that a good number of the athletes died from heart attacks.[68] A fair number of those athletes who managed to survive did so only by quickly commencing vigorous exercise to thin their blood temporarily until the crisis passed.

This drug is very costly. For example, just to prevent temporary blood loss due to surgery, the estimated cost for a 70-kilogram patient is between $1,000 and $3,000. The continuing cost for a person undergoing dialysis or for a premature infant is much higher. Worldwide sales of EPO run about $4 billion.

One of the reasons for the high costs is that some adult patients and nearly all premature infants need to be given very high levels of EPO because the body immediately excretes it unless the patient has a naturally occurring binding factor. Most people have this binding factor that keeps the EPO from being excreted in the urine, but many do not, especially those with immature renal systems such as infants and children.

Gisela Clemons, a scientist at the Lawrence Berkeley National Laboratory, discovered a protein-binding factor that binds EPO in the body so that it is not excreted, dramatically decreasing the dosage of EPO required for those who lack the binding protein. She was hopeful that this binding protein would save the lives of many premature infants around the world. On April 29, 1997, the Patent and Trademark Office awarded U.S. Patent 5,625,035 for this drug.

Well before the patent was issued, the Berkeley lab offered it to drug companies, including Amgen, which owns most of the patents on EPO. Amgen was not interested. Although the company denies it, many of the people involved think Amgen's disinterest was financially motivated.

With this binding factor, those people who lack the binding protein would need much less EPO per dose. This new innovation would cut the demand for

EPO across all markets by an estimated 50 percent. Of course, Amgen had no reason to want to develop this product. Other drug companies were not interested because they would have to combine the binding protein with EPO, and most of the rights to EPO were in the hands of Amgen.

Shortly after Luehrmann wrote of the difficulties in getting this lifesaving invention out to the public she resigned from her job at the Berkeley lab, saying that her job was made untenable after the binding-protein disclosure. Perhaps not coincidentally, according to Luehrmann, Amgen makes substantial donations to the Berkeley Lab's research programs.

Amgen did not invent EPO. Other scientists had isolated it from human urine well before Amgen even existed. One doctor even tried treating three anemic patients with it. Amgen won a patent for its method of producing EPO. Its technique is to clone the human gene governing the production of EPO, and then implant the gene into hamster ovary cells, which then make the protein.

Transkaryotic Therapies Inc., presently a tiny company, has developed an alternative technology that could allow it to circumvent patents on drugs produced by genetic engineering, beginning with EPO. All human cells have a complete set of genes; even though the gene for producing EPO is inactive except in kidney cells, "Transkaryotic inserts an 'on switch' into human cells grown in culture, activating the gene for EPO. The switch is surrounded by genetic sequences that match those found upstream of the EPO gene on the chromosome. That guides the switch to the spot where it will turn on the EPO gene and not any other gene."[69]

The legal system is now deciding if Transkaryotic can sell its own version of the drug. A federal court ruled in favor of Amgen, but appeals are certain. Again, the struggle over ownership of intellectual property rights will consume enormous time and resources. The bemused judge noted in his decision the zeal with which reporters and financial analysts would seize upon the slightest details to try to get a feel for the way the trial would end.[70]

The academic world has also entered into the fray over EPO. Columbia University owns a patent that covers a technique that allows animal cells to be used to manufacture proteins used as drugs. The patent has been used to produce many of the biotechnology industry's best-selling drugs, such as EPO and Immunex Corporation's Enbrel, used for rheumatoid arthritis. The school gets a royalty of 1 percent of the sales of such drugs and has collected $280 million since its patent was granted in 1983.

As the sales of the drugs continue to grow, so too will the royalties. Unfortunately for Columbia, the patent was scheduled to expire in August 2000. As a last-ditch effort, the university enlisted Judd Gregg, a Republican senator from New Hampshire and member of the Columbia class of 1969, to propose an amendment to an agricultural spending bill in an attempt to allow the school to continue to collect royalties.

Of course, the pharmaceutical companies jumped into the fray, opposing this legislation to create an extension to an intellectual property right.[71] While these companies rapaciously exploit intellectual property rights, they naturally resent "technology players [that] have positioned themselves as 'tollgates' on the road to drug discovery."[72]

The Tragic Consequences of Litigation

Royalties are not the only obstacles to developing better medicines. Wasteful litigation creates formidable roadblocks. A particularly poignant example concerns the struggle between Baxter International, the global medical-products company, and a small biotechnology company called CellPro Inc. According to a detailed report in the *Wall Street Journal,* CellPro had developed a very promising technology designed to combat a particularly lethal form of advanced leukemia. Baxter had been developing a competing system for the experimental procedure.

Baxter challenged CellPro's patent in court. Initially, a jury ruled against Baxter, but the judge reversed the decision. After losing in court, CellPro Inc. canceled its clinical trials. The company reported that because its five-year legal battle with its much larger corporate rival had exhausted its finances, it found itself forced to seek chapter 11 bankruptcy protection. Its product lies abandoned.

As a result, a technique that the experts consider far superior to Baxter's was not tested. In addition, Baxter sold the division that made its version of the product to another company, which caused delays in the testing of its product.

In other words, the system of intellectual property first denied patients access to CellPro's supposedly superior product, and then to add insult to injury it held up the testing of Baxter's inferior version. The upshot of this case is, as the *Wall Street Journal* noted, "hundreds of otherwise-doomed cancer patients have lost a shot at a last-ditch experimental treatment that might offer a ray of hope." Hillard Lazarus, a cancer specialist at Cleveland's Case Western Medical Center, says his institution, like several others, has turned away some dying patients who would have qualified for CellPro's trials. "We're fighting a battle with one arm behind our back," he says. Moreover, the product that Baxter was promoting was inferior to CellPro's. "The operable word here is avarice," Dr. Lazarus says. "What did these companies accomplish? CellPro is essentially dead. We lost a souped-up race car and now we're left with a horse and buggy."[73]

Had the time and resources spent in litigation been directed toward research and development, an even better technology might well have been on the market for some time now. Similarly, had the companies' researchers been free to cooperate, eliminating duplication of efforts, no doubt the technology would be still more advanced.

Litigation disrupted an even more vital line of research: the development of a mouse with a human immune system. Scientists had been producing clones of antibodies in mice. Although these antibodies had the potential of being highly targeted drugs, these so-called monoclonals did not fulfill their promise because of immune responses. GenPharm International set out to create a mouse with a human immune system that would be capable of manufacturing antibodies that would not set off an immune response.

The company announced its intentions, but in February 1994, just as it was about to issue stock that would fund its research, another firm, Cell Genesys, sued GenPharm, charging it with having stolen a trade secret for inactivating a mouse gene. A *Scientific American* article described the consequences of the suit:

> The company fired everyone but a skeleton staff and began to try to find a buyer, but the lawsuit stood in the way. GenPharm, which at one point had a mere $15,000 in cash and owed large sums to lawyers and banks, had to survive on barter; it tended mouse cages for another biotechnology concern in exchange for a sliver of laboratory space. It held a fire sale to get rid of furniture, laboratory equipment and patents. A professor of developmental biology from Stanford University came down and inspected a surgical microscope as a possible toy for his kid.[74]

Two weeks before the trial in early 1997, Cell Genesys dropped its suit. To meet its need for cash, GenPharm worked out an agreement with Cell Genesys and a partner company that would provide nearly $40 million for a cross-licensing agreement. Nils Lonberg, the lead scientist for GenPharm, summed up the dispute: "The final story is not that we prevailed but that [Cell Genesys] actually succeeded in its strategy. It was able to use litigation to capture a technology."[75]

The Callous Neglect of the Less Affluent Regions

Corporate science narrows its focus to those areas of research that will produce the most profit. As a result, the needs of the poor will not be a high priority.

Recall the earlier discussion about the heritage of medicinal information from the people who traditionally lived in the tropical rain forests, regions with the highest degree of biodiversity on the earth. Even when the indigenous people of a region had discovered the medicinal properties of a plant, more often than not they receive no benefit in the form of intellectual property rights. While intellectual property rights for corporate science are strong, to say the least, they are nonexistent for owners of traditional knowledge.[76]

Instead, multinational corporations visit indigenous people, take traditional information about the medicinal properties of plants, and then patent it as if it were their own invention. Then, according to the current legal structure, once their knowledge becomes a commercial product, the people who live in these societies are obligated to pay for the fruits of the information that they first developed. Not surprisingly, those who find themselves at the short end of intellectual property law have difficulty in comprehending the justification of their predicament. To add to the absurdity of this situation, the multinational pharmaceutical companies do almost nothing to address the serious health problems of these people who contributed mightily to the medicinal knowledge of modern society.

The response of the pharmaceutical companies to the AIDS epidemic in the poor parts of the world is even more appalling. In 1997, the gross domestic product per person living with AIDS ranges from $616,210,735 in Japan to $9,553,702 in the United States all the way down to a mere $2,294 in Mozambique.[77] Meanwhile, the disease spreads, threatening a generation of young Africans, undermining the economy, and making the poor people of the continent even less able to afford the market price for the drugs necessary for their survival. A recent report by the World Health Organization estimated that 35.8 percent of the population of Botswana is HIV-positive. In Botswana, life expectancy has fallen from over 70 years to less than 40 years because of AIDS. If present trends continue, by 2010 conditions will worsen even further: life expectancy will only be 29 in Botswana, 30 in Swaziland, and 33 in Namibia and Zimbabwe.[78]

In 1999, under intense pressure from AIDS activists who were embarrassing Vice President Al Gore, the putative Democratic candidate for president at the time, the government of the United States stopped pressuring the South African government to pay the full price for AIDS drugs. Later, under the continuing glare of unfavorable publicity, the pharmaceutical companies agreed to lower the cost of AIDS drugs in Africa for a few countries.

Then the companies sued the South African government, which was taking advantage of a law that enabled it to import generic drugs. They finally withdrew that suit, but only after extracting key concessions that weakened South Africa's ability to import generic drugs in the future. This ongoing fight against affordable health care constitutes the real drug war of today.

The pharmaceutical companies' stubborn insistence on charging the full price for AIDS drugs had little to do with the loss of revenue from the African market. Although the profits per unit of medicine is as high in Africa as in the rest of the world, relatively few people in the poor nations of Africa can afford to purchase these drugs. As a result, profits from the African nations are miniscule by corporate standards.

The real concern of the pharmaceutical companies was that concessions in Africa would lead to demands for lower prices in the larger, more affluent markets. Any sign of weakness in Africa might open the floodgates for others to demand lower drug prices, especially in the United States, where the pharmaceutical companies charge considerably more than they do in other advanced industrial nations. A *Wall Street Journal* article laid bare the thinking in the pharmaceutical companies:

> According to people familiar with the situation, executives were worried that slashing prices would reveal the industry's large profit margins. For example, after a series of orchestrated attacks by activists, Pfizer Inc. in December agreed to give away its antifungal drug Diflucan to certain AIDS patients in South Africa, rather than lower its price and reveal the scope of its profit.[79]

The pharmaceutical companies, which are among the most profitable businesses in the world, defend their practices in terms of the rights flowing from their ownership of intellectual property. Of course, the value of property in general always depends on the legal structure. For example, within the context of a traditional industrial setting, the legal rights of workers to form effective unions affects the wealth and income of factory owners relative to their employees. While economists, and to some extent the public at large, have given some consideration to the implications of the legal rights of unions, the broader social impact of the strengthening grip of intellectual property rights has gone virtually unnoticed. I will indicate why intellectual property rights represent a particularly pernicious mutation of the concept of property—one that is thoroughly detrimental to the social and economic health of a community.

Shamelessly, modern drug companies probably devote more resources to developing medicines for pets in the affluent world than for drugs to help people in the tropical world. Already, Novartis makes a drug to treat separation anxiety in dogs. Pfizer has developed a medicine to treat dog Alzheimer's.[80] Pfizer is expanding its research on pet medicine by about 15 percent per year.[81]

The drug companies are not unmindful of places such as Africa. When an epidemic strikes that impoverished part of the world, they can use the opportunity to experiment hastily with drugs that have not yet been tested on humans. If their experiment works, they can save many millions of dollars. A thriving industry has emerged devoted to finding places to do human research in less developed countries that provide virtually no oversight for such activities.[82]

If one of these experiments leads to a tragic outcome, nobody outside of the affected area is likely to know. In one exceptional case, Pfizer's activities came to the attention of the press. Not only did the affected population suffer harm from the "treatment," but others lost the opportunity for treatment because the experiment absorbed scarce medical resources, such as hospital beds.[83]

To turn a profit by catering to the rich while overlooking the grievous medical issues that afflict the poor is standard operating procedure for the industry. Chris Scott, who until recently oversaw industry collaborations at Stanford's medical school, observed, "Show me an industry-sponsored research project on schistosomiasis—a liver parasite that afflicts people in the Third World—or malaria or river blindness or dengue fever."[84] Ken Silverstein, an excellent investigative journalist, reported:

> Only 1 percent of all new medicines brought to market by multinational pharmaceutical companies between 1975 and 1997 were designed specifically to treat tropical diseases plaguing the Third World. In numbers, that means thirteen out of 1,223 medications. Only four of those thirteen resulted from research by the industry that was designed specifically to combat tropical ailments. The others, according to a study by the French group Doctors Without Borders, were either updated versions of existing drugs, products of military research, accidental discoveries made during veterinary research or, in one case, a medical breakthrough in China.[85]

Similarly, the United Nations Development Program notes:

> Of the annual health-related research and development worldwide, only 0.2% goes for pneumonia, diarrheal diseases and tuberculosis—yet these account for 18% of the global disease burden. In the United States between 1981 and 1991, less than 5% of drugs introduced by the top 25 companies were therapeutic advances. Some 70% of drugs with therapeutic gain were produced with government involvement. Vaccines are the most cost-effective technologies known in health care, preventing illness in a one-time dose. But they generate smaller profits and have higher potential liabilities than treatments used repeatedly. As a result a consortium of US pharmaceutical companies has united to develop antiviral agents against HIV, but not to produce a vaccine against AIDS.[86]

To make matters worse for people in the poorer regions of the world, the U.S. government, along with the World Bank and the International Monetary Fund, consistently pressure less powerful nations to reduce their own public investment, forcing them to rely even more on the market for their medicines. The market, however, has little interest in the diseases that plague Africa. The following three case studies illustrate this tragic indifference.

Sleeping Sickness, Malaria, Schistosomiasis, and the Ebola Virus

I want to continue with the theme that the market displays a callous disregard for those without adequate funds to satisfy the corporate lust for profits. Consider the example of sleeping sickness, a disease that infects about 300,000 people each year. The only available treatment for this disease is melarsoprol. This drug, "invented almost 70 years ago, is melarsen oxide dissolved in propylene glycol—literally, arsenic in antifreeze."[87] It is so toxic that it kills 5 percent of those treated with it. To make matters worse, a strain of melarsoprol-resistant sleeping sickness is now spreading.

African doctors now have no other medicine available to them to treat advanced sleeping sickness. Until July 1999, some patients could use a superior medicine, eflornithine, known as DFMO and sold under the name Ornidyl. This drug is the best-known treatment for those melarsoprol-resistant patients.

The industry originally developed eflornithine as an anticancer drug. Only by chance did it learn about its usefulness against sleeping sickness, but "it proved so spectacular at pulling people out of their final comas that it was nicknamed 'the Resurrection Drug.'" The manufacturer, an American subsidiary of Aventis, abandoned eflornithine in 1995 after it proved ineffective against cancer. Presumably, the African market was too small to satisfy the high rate of return that the pharmaceutical companies expect, even though the drug costs about $210 per course of treatment. Recently, however, interest in the precursor chemical to eflornithine has suddenly soared because it might prevent the growth of facial hair in women.[88]

The treatment of malaria is as scandalous as that of sleeping sickness. Malaria kills more than a million people a year, mostly children. The human cost of malaria is incalculable. A mere $8 devoted to prevention is sufficient to add one year of healthy life to an African. One study found that malaria has cut the economic growth rate in Africa by one full percentage point. While that effect might not seem like much, let me put it in perspective. Had the disease been eliminated in 1965, Africa's gross domestic product would have been one-third higher in 1995 than it actually was.[89] Not surprisingly, the toll that malaria imposes on the people of Africa varies considerably from country to country. In those countries where the gross domestic product is increasing most rapidly, where income is more equally distributed, and where people have the most access to rural health care, the morbidity associated with malaria is reduced.[90]

Even when the people of the Southern Hemisphere made progress in fighting their indigenous diseases, they got no support whatsoever from the giant pharmaceutical companies, until the recent popular movements forced these corporations to begin to make some cosmetic gestures. Consider the case of the pharmaceutical companies apparent lack of concern with schistosomiasis. According to a December 1998 press release of the World Health Organization, over 200 million people worldwide suffer from schistosomiasis—an infection that, if not controlled, can cause the development of liver and urinary tract disease and cancer of the bladder. Over 20 million people are chronic sufferers, most of them living in the 48 African countries where the disease is endemic.

While studying this disease in 1964, Akilu Lemma, of Addis Ababa University, found that a local endod berry had the capacity to control the parasite, with no apparent adverse side effects. Scientists from the National Research Development Corporation of London offered to collaborate with him. He sent them samples, but without communicating with him any more, they took out a patent. He later shamed them into donating it to his foundation.

To make the medicine more available, Lemma had to get the blessing of the World Health Organization, but that agency would not vouch for its safety without tests. No pharmaceutical company would support such a test since they could not get rich off a plant that is so abundant. So instead, the World Health Organization recommended a chemical to kill mollusks marketed by Bayer for $27,000 per ton compared to his chemical, which cost $1,000.[91]

The Ebola virus does hideous things to a human body, causing the internal organs to disintegrate. Fearing that, with the massive number of intercontinental travelers and the possibility of mutations, the virus could possibly threaten to create an epidemic in the United States, the Centers for Disease Control and Prevention developed an experimental vaccine that could protect monkeys against Ebola. Since the disease is presently confined to impoverished African nations, the pharmaceutical industry has no interest in following through and creating a safe vaccine for human use. "There's no market for this," says Gordon Douglas, the recently retired chief of vaccines for Merck, one of the world's leading vaccine developers.[92]

Not surprisingly, one of the most dramatic breakthroughs in tropical medicine recently came from Cuba, a country where the influence of the United States is relatively weak. A British company, SmithKline Beecham, recently signed a license agreement with the government of Cuba to market a new vaccine for meningitis B, one of the world's principal killers of children and adolescents, affecting some 500,000 and causing the death of about 50,000 around the globe each year. Cuba's state-owned Finlay Institute developed the drug in the 1980s after the island experienced an epidemic of the disease.[93]

The Cuban pharmaceutical industry has made breakthroughs in other areas. For example, it has developed a cancer therapy apparently capable of increasing the average survival time by 200 percent. A new antioxidant is also promising.[94]

How can we explain why a small island, unable to acquire many vital goods because of a longstanding embargo by the United States, is able to develop breakthrough drugs, while the multinational firms with their huge resources protected by the umbrella of intellectual property rights manage to create only a handful of drugs applicable to tropical diseases? Perhaps if global warming continues to move tropical diseases into more temperate zones the major pharmaceutical corporations might display more interest in tropical diseases.

Even in the absence of global warming, the failure to address the health of poor people in less developed countries can bring harm to people in the more affluent areas. Sick, undernourished people with crippled immune systems provide ideal breeding grounds for new strains of disease-causing agents. With the upsurge in globalization, the spread of such lethal agents to the advanced countries becomes more likely.

A Real Headache

One of the most ominous aspects of the broadening of the scope of intellectual property rights is the extension of patents to human genes. Over and above the ethical questions regarding the patenting of human life, this form of intellectual property raises serious questions about the future of medical science, as well as the distribution of wealth and income.

The University of Rochester recently won a patent that covers the use of all drugs that inhibit a form of the enzyme, cyclooxygenase, known as COX-2. These drugs, often described as super aspirin, relieve inflammation apparently without the risk of ulcers and other gastrointestinal problems. Soon after receiving the patent, the school filed suit against Pharmacia, claiming a 10 percent royalty on the company's total sales of Celebrex, a popular painkiller. The university also sued Pfizer, which comarkets Celebrex with Pharmacia. It also plans to collect royalties on another profitable painkiller, Vioxx, which Merck markets.[95]

Merck and Pfizer were expecting to sell about $4 billion worth of the painkillers in 2000. Each side now has more than 4,000 representatives visiting doctors to promote their drugs.[96]

The history of this medicine, like that of most technologies, is tangled. Two decades ago, Philip Needleman, then a researcher at Washington University, in St. Louis, and his co-workers postulated the existence of two cyclooxygenase enzymes, COX-1 and COX-2. By 1990, Dr. Needleman, then chief scientific officer at Pharmacia, had guessed that the COX-2 enzyme plays a critical role in inflammation. By 1992, three other groups, including one at Rochester, had confirmed the existence of the enzymes by describing the genes that control their production. Although Rochester won the patent, the competing teams at UCLA and Brigham Young University claim that their work was fundamental.[97]

Whether UCLA, Brigham Young, or Rochester deserved the patent is beside the point. More important is the idea that the granting of a patent on a bodily substance permits the owner to demand royalties from any company that produces a medicine that targets that substance. A recent article in Massachusetts Institute of Technology's *Technology Review* pointed out the stakes in allowing such patents:

> The number of human genes is likely somewhere between 50,000 and 150,000 (geneticists are still debating the actual number). Of those, about 1,000 have already been patented, and applications on thousands of others await approval. . . . All existing drugs on the market act on only about 400 distinct targets in the body; these are the critical enzymes and pathways that can be addressed in treating various diseases. And scientists' best guess is that there may be only 5,000 "druggable" targets overall. As they reveal these targets, companies and universities are laying claim to what amounts to the ultimate biological monopoly, patents that comprehensively cover not only a gene and the protein it makes, but also any "method of treatment" that specifically targets them. Already, Pfizer has a patent in Europe covering any drug that acts via PDE5, the protein targeted by Viagra.[98]

Mark Boshar, chief patent counsel for Millennium Pharmaceuticals, identified claims for such "methods of treatment" as the "crux" of his company's patent strategy.[99] By the middle of the year 2000, Millennium already had a staggering 1,500 applications pending at the U.S. Patent and Trademark Office.[100] In effect, these companies position themselves as tollgates,[101] but, as anyone who has driven through a tollgate knows, they can generate terrible traffic jams.

As the companies grow rich collecting these tolls, they will accumulate more political influence, insulating them from any legal challenges to their intellectual property rights. "This is a land grab of historic proportions," said Michael S. Watson, a leader of the American College of Medical Genetics, the national professional organization of genetic practitioners. "It's as if somebody just discovered English and allowed the alphabet to be patented."[102]

Creating Intellectual Property in Genes

A company named Incyte leads in the human gene patent race with 353 U.S. patents issued, followed by Human Genome Sciences with 114, and SmithKline Beecham with 60. The next three highest holders form an interesting group: the U.S. government (49), the University of California (46), and Massachusetts General Hospital in Boston (45).[103] The presence of the University of California on the list serves as a reminder of how similar business and the leading academic institutions have become.

Corporate science uses two main techniques to develop genetic patents. First, companies sift through the DNA of a specific population group looking for genetic irregularities associated with a particular condition. Second, applying technology developed by the U.S. government, corporations use powerful computer-driven DNA-sequencing machines to decode as much of the human genome as possible. This technique depends more on the brute force of technology than on the creative and innovative research associated with pure, or even applied, science.

A recent *Wall Street Journal* article describes the process that Incyte uses, built around a "gene-finding assembly line" with a series of these DNA-sequencers. "At the end of his assembly line stood patent writers, who cranked out dozens of sketchy applications each week as placeholders, describing only a fragment of a gene until Incyte's scientists could decode its entire length." "Three teams, including 60 Ph.D.s, sifted and categorized the genes as best they could. One team ranked them according to probable medical value, one scoured the literature for clues about their functions, and one wrote the patent applications, complete with computer-generated comparisons of the genes with others in the database. In a sprawling basement office, computer printers spewed out paper; some patent filings were 12,000 pages long."[104]

Incyte now holds U.S. patents on more than 350 genes has about 6,500 applications pending and is applying for about 100 more a week.[105] Any one of these patents is potentially worth millions, even billions of dollars. To put this operation into perspective, recall that the Patent and Trademark Office gives its examiners only 80 hours to determine the validity of a patent, or 150 pages per hour for a 12,000-page patent application.

A competitor of Incyte, Human Genome Sciences, has already received more than 100 human gene patents and has 7,500 patent applications pending. In June 1995, the company applied for a patent for one of those genes, which produces a protein receptor that the company identified as "HDGNR10." The company was awarded a patent on the gene in February 2000.

A number of researchers had independently discovered that HIV enters cells by grabbing hold of certain protein receptors that act like handles for the AIDS virus on the cell surface. The most important of these is called the CVCR5 receptor. These scientists isolated the protein and found the gene that produced it. The gene and the protein it produces are identical to those previously isolated by Human Genome Sciences—something that the academic researchers could not possibly have known because the company's patent application, although filed, was still not publicly disclosed.[106]

Although a patent applicant is supposed to demonstrate some utility for an invention, so far the corporations that are patenting gene fragments have "gambled on a legal loophole; they proposed generic and often frivolous uses—as forensic probes, even cattle feed."[107] The Patent and Trademark Office is proposing to limit this loophole, but presumably they will not revoke existing patents.

So here is Human Genome Sciences. The company did no work on curing AIDS. It merely identified a gene and filed for a patent. Except for its patent application, it did not even publicize its work. Companies that now begin to develop a cure for AIDS by trying to disable the CVCR5 receptor may find themselves having to pay royalties to the holder of the patent, siphoning off scarce funds that could contribute to the cure for the disease.

Two points are very relevant in assessing the intellectual property claims of the holders of patents in human genes. In the first place, these companies profit considerably by the publicly owned Human Genome Project, which makes its findings public. Second, the key technology in the rapid decoding of these genes is the genetic sequencer, which was also developed using public funds. To the extent that the government funded this technology, it can claim certain rights for itself.

The government did not just fund the research to develop the sequencers. It also provided the initial market that allowed the manufacturer to earn profits before the industry matured. Until recently, the government and government-funded scientists were the largest purchasers of these sequencers, having bought about 6,000 of them over the last decade. In addition, under a 1980 law, the government is owed a discount on the products of federally funded research. As of February 2000, federal officials, led by the inspector general's office in the

Department of Health and Human Services, were investigating whether the government overpaid millions of dollars to purchase the machines, the latest version of which sells for $300,000.[108]

PE Biosystems is the main manufacturer of these machines, with an estimated 92 percent of the market. It also owns Celera Genomics, which runs 300 PE-made sequencers and some of the world's fastest computers 24 hours a day in an effort to be the first to map the entire human genome. In a congressional hearing in spring 1991, James Watson, the codiscoverer of DNA, said that "virtually any monkey" could do what Celera was doing since the real work was done by the computers.[109] The computers, incidentally, also were developed with public money.

Naturally, PE Biosystems denies that the government funds played a role in the development of the technology, although the inventors' laboratory notebooks, which have been submitted under subpoena to federal investigators, contain a sustained lament by the researchers about their inability to get the machine to function correctly during the crucial period just before the start of federal funding.[110]

The stakes in the ownership of these gene patents are enormous. In March 2000, President Clinton and Prime Minister Tony Blair of Britain said that the sequence of the human genome should be made freely available to all researchers. In response, the stocks of biotechnology companies fell sharply. During the following day, Incyte's stock fell $53.50, to $143.50, Human Genome's stock sunk $29.04875, to $123.51625, and Celera's slumped $39.75, to $149.25.[111]

The movements in the stock market were reflecting potential changes in the extent to which costs could be heaped on pharmaceutical companies, which would then pass them on to consumers. Pharmaceutical companies, which revel in huge profits from their own intellectual property, complain that royalties paid to holders of patents on genes, research mice, and other tools can total 12 percent to 14 percent for a single drug, making some products uneconomical to produce.[112] They find such intellectual property rights inconvenient, although they are unwavering in their belief that those rights that they hold themselves are in the public interest.

Despite the questionable practice of patenting genes, holders of these intellectual property rights continue to expand their grasp. In one of their more outrageous efforts, some corporations now are attempting to extend their patent claims to include ownership of any "computer-readable medium having recorded thereon the nucleotide sequence."[113] *Scientific American* commentator Gary Stix observed that such patents would have potentially far-reaching consequences. Merely accessing the genetic sequence over the Internet could constitute an infringement.[114] Law professor Rebecca Eisenberg observed that such claims would violate the entire rationale of the patent system:

> If the Federal Circuit steps back from the momentum of its recent decisions expanding the boundaries of the patent system, it should not be persuaded that information stored in computer-readable medium is patentable. Patent claims to information—even useful information—represent a fundamental departure from the traditional patent bargain. That bargain has always called for free disclosure of information to the public at the outset of the patent term in exchange for exclusionary rights in particular tangible applications until the patent expires.[115]

Genetic Detective Work

Rather than mechanically sequencing genes by brute force of technology, the second technique for hunting down genetic pathways follows what seems to be more in the tradition of pure science. In the end, however, the dominant financial motives pervert the science. In addition, this technique raises serious concerns about invasion of privacy.

This approach entails identifying a group of people with a specific feature, then searching for the genetic cause. For example, in 1990, after years of research, a team led by Dr. Mary-Claire King demonstrated that some women affected with breast cancer in high-risk families shared genetic markers in common in a region of the long end of chromosome 17. Dr. King identified and named this gene BRCA1 (for breast cancer). In 1994, a group in the United Kingdom identified another set of women in high-risk families who shared genetic markers on chromosome 13 and named this gene BRCA2.[116]

Both BRCA1 and BRCA2 are thought to be tumor-suppressor genes. When they function properly, they inhibit the uncontrolled growth of cells. A person who inherits one mutated copy of the gene, from either parent, is at a higher risk for cancer. If radiation, chemical carcinogens, diet, or some other factor, either genetic or environmental, damages the remaining good copy of the properly functioning gene in a cell, it loses this source of control and has taken a step toward becoming tumorous.

A woman with a mutated BRCA1 or BRCA2 gene has an estimated lifetime risk of contracting breast cancer of approximately 70 to 85 percent, compared with a risk of roughly 10 percent in the general population. BRCA1 mutations also confer a heightened risk of ovarian cancer. Patients in high-risk families who have mutated BRCA1 genes have an estimated 28 to 40 percent lifetime risk of being diagnosed with ovarian cancer, compared with the roughly 2 percent lifetime risk of this disease in the general population. Specific mutations occur with high frequencies within particular populations, including those of Eastern European Jewish (Ashkenazic), Swedish, Icelandic, and Finnish ancestry.[117]

Once the basic scientific research in isolating the breast cancer gene was complete, a classic patent race began to decode the gene. Had these researchers pooled their energies, they might have been able to accomplish wonders. Instead they merely duplicated one another's efforts in a competition for the huge profits that awaited the winner.

In the end, the winner was Mark Skolnick of University of Utah, who set up Myriad Genetics, Inc. The U.S. Patent and Trademark Office awarded Myriad Genetics patents for the BRCA1 and BRCA2 genes, as well as several other patents related to these genes.

As a result of its gene patent, Myriad can charge a royalty to people doing research with the genes that it "owns." If a test looks for a particular, known mutation, such as 185delAG, testing is relatively cheap, with direct charges of approximately $350. If, however, the test needs to examine the full length of the gene for mutations, the patient's entire gene must be sequenced. Such tests are extraordinarily expensive. OncorMed charges $1,995 for full BRCA1 testing and $2,100 for full BRCA1 and 2 testing. Myriad charges $2,400 for full sequencing of BRCA1 and 2.[118] According to Myriad's Web site, growth of its genetic testing revenues has averaged a fantastic 25 percent quarter to quarter.

In early 2000, Sir Walter Bodmer, a geneticist with the Imperial Cancer Research Fund and head of Oxford University's Hertford College, called on the British government to challenge the patent application, charging that Myriad had done nothing to merit exclusive rights. As Andrew Read, chairman of the British Society of Human Genetics, observed, a question exists as to whether the gene was discovered by Myriad or the London-based Institute of Cancer Research. More important, "there was no inventive step. Any Ph.D. student could write down how to do it." In effect, according to Read, Myriad merely put the last brick in a wall to which many people had contributed.[119]

As a result, researchers report cutting back on their efforts to understand the gene for fear of infringing on the patent rights of Myriad Genetics. Arupa Ganguly, research director of the University of Pennsylvania's genetics laboratory, laments, "Patients still keep coming, begging me to do the research, but I cannot anymore."[120]

One survey taken in conjunction with the American Association for Clinical Chemistry and the Association of Molecular Pathologists attempted to estimate the impact of patenting on genetic diagnostic testing on clinical genetic services. The survey found that about 25 percent of the labs contacted had decided not to develop or perform a test because of a patent.[121] In addition to the royalties that laboratories pay to do research, any company producing a medicine that targets a patented gene has to pay an additional royalty.[122]

Primitive Accumulation Again

The process of identifying genetic problems is much simpler when the search is confined to a relatively small group of people who live in remote areas and who have been inbreeding over a long period of time. Researchers can more easily narrow down their search for the trigger gene since the genetic variations are minimal compared to a larger population.

For example, in a typical population, two individuals differ from one another roughly once every 300 DNA bases of a total of 3.3 billion bases, for about a 11 million places in their genetic code. Because the genetic diversity among people is so limited within these remote, inbred populations, the task of pinpointing a genetic defect becomes far simpler. Rather than having to search a whole haystack to find the proverbial needle, researchers need only scour through a small pile of straw.

In one case, scientists noticed that asthma was common among the 301 people who inhabited Tristan da Cunha, a tiny island roughly halfway between Cape Town and Buenos Aires, where 3 of the original 5 ancestors supposedly suffered from asthma. Inbreeding is so common that on average 2 siblings are cousins according to 50 different pathways. A team of scientists with the University of Toronto Genetics of Asthma Project, led by respiralogist Noe Zamel, descended on the island and took blood samples from 282 inhabitants.[123] Because of the relatively homogeneous gene pool, the scientists expected to be able to identify a gene that they believed caused the asthma.

The researchers still had to isolate the specific gene associated with asthma. The plan was to return with their blood samples to Mount Sinai Hospital in Toronto, where geneticist Kathy Siminovitch was to extract the DNA and look for the defective genes that caused the ailment. A new California company later named Axys Pharmaceuticals Inc. read about the project. It formed a partnership with Mount Sinai, providing the money and technology to do the robotic part of gene mapping of the DNA that Sinai had. Soon after, Axys entered into an agreement with German chemical company giant Boehringer Ingelheim, which pledged more than $30 million to the project. Boehringer's funding went directly to Axys, leaving Mount Sinai out of the loop. The corporations sent Dr. Zamel to China, Brazil, and back to Tristan da Cunha. Mount Sinai also collected blood samples in Canada, but it never even got to see the scientific results.[124]

Just as the rush for profit showed no regard for the group at Mount Sinai Hospital, it took no account of the people of Tristan da Cunha. They were seen as little more than raw material, the value of which was just waiting to be extracted. In the words of Tim Harris, vice-president of the company that initially contracted for the research: "The Tristan population was the tool that opened the box."[125]

Axys fared poorly in its endeavor to develop a drug. The company discontinued clinical studies of its lead asthma treatment because it spurred coughing and wheezing in some people. It closed its San Diego research facility. Axys still remains eligible for a series of preclinical and clinical milestone payments from Rhone-Poulenc, which took over the research on one drug that might treat asthma and other inflammatory disorders.

Defenders of intellectual property might read the wrong moral into the story of the search for the asthma gene. Yes, the search for genetic information may be expensive. It can also be risky. The payoffs, however, can be enormous, making the risks well worth taking.

We need not shed too many tears over Axys's setback in its quest for an asthma drug. Although its stock price fell below $3 per share in 1999, hope springs eternal in the investment world. Perhaps carried along by a powerful speculative wave in biotechnology stocks, the company's shares rebounded back to almost $15 in the early months of the year 2000, only to retreat back down below $4 by the end of the year.

Others have followed in this path of mining the genetic pool of isolated populations from Newfoundland to remote regions of China. None have been more ambitious than the venture of Kari Stefansson, to which I will now turn.

DeCoding Iceland's Genes

Kari Stefansson, an Icelandic native, who was formerly a professor of neurology and neurosciences at Harvard University, formed a company, deCode, that signed an agreement with the government of Iceland to get exclusive access to the genetic pool of that remote island.

Stefansson insists that Iceland is an ideal site for this type of research.[126] Iceland has kept meticulous genealogical records for centuries. Its genealogical database includes nearly 75 percent of all of the Icelandic people who have ever lived. In addition, Iceland has kept detailed records concerning almost every major illness since 1915.[127]

Iceland's fascination with genealogy runs so deep that its newspapers publish regular columns on the subject and its university employs a professor of genealogy. The people of Iceland even know now that a sixteenth-century Icelandic cleric named Einar passed on a defective BRCA2 gene that is responsible for much of the hereditary breast cancer in Iceland today.

Stefansson claims, "The native genome of Iceland offers a powerful and rare resource in genomic research—a relatively homogeneous population."[128] Three scientists from the University of Iceland recently contested this assertion. After studying biological samples from a group of Icelanders, they concluded, "Icelanders are among the most variable Europeans at the mtDNA [mitochondrial DNA] level."[129]

DeCode won the exclusive rights to this valuable database, despite the vigorous opposition of the Icelandic scientific and medical establishment. Its chances in the parliament were not harmed by the fact that, according to published reports, deCode admittedly offered money to all political parties that so requested. In addition, deCode allegedly purchased a building to serve as its headquarters from a company called Hans Petersen, which is owned by the family of the prime minister's wife. She is also a shareholder and a member of the board of directors.[130] Notwithstanding the strong opposition of the national medical society, the company is signing exclusive agreements with individual physicians, through whom it gains access to patients, which prevents other geneticists and or biotech companies from competing with deCode.

If deCode's activities lead to a medical breakthrough, we can rejoice, but the question remains whether or not its approach is the best way. Of course, deCode's market is not only the discovery of medicine. It intends to sell its database to companies that market medical insurance. If an insurance company knows that a genetic marker indicates a high probability of a costly medical condition, it can reject carriers of that gene, thus reducing its average costs. This strategy will throw the costs of treating those most in need of treatment onto the individuals, their families, or the state.

Although deCode is still in its infancy, Hoffman-La Roche has already offered deCode $200 million over five years to identify genes for 12 specific diseases: 4 of the brain (schizophrenia, anxiety disorder, Alzheimer's, and manic-depressive disorder), 4 cardiovascular diseases (heart attack, peripheral vascular disease, high blood pressure, and stroke), and 4 other diseases (osteoarthritis, osteoporosis, non-insulin-dependent diabetes, and emphysema). The risk for Hoffman-La Roche may not be great since it holds a majority stake in deCode, which coincidentally filed a $200 million initial public offering in March 2000.

The idea of tapping the Icelandic gene pool was no doubt clever. Certainly, society should acknowledge such contributions. Historically, the open acknowledgement of the scientific community was a sufficient reward to inspire scientists to greatness. But not for Kari Stefansson, who had spent time at the University of Chicago before relocating to Harvard. He arrogantly proclaimed, "I did not spend 15 years at Milton Friedman's university for nothing . . . this is my intellectual property."[131]

Milton Friedman's free-market views notwithstanding, monetary rewards of hundreds of millions of dollars seem excessive. To put this affair into perspective, the head of an Icelandic bioethics organization observed: "The company is expected to reach a value twice as high as the country's yearly revenue."[132]

Unfortunately, Milton Friedman's ideals cut both ways. Medical science has long relied on the voluntary cooperation of people who wanted to contribute to the advancement of science. Once science becomes a predominantly commercial venture, people become reluctant to make voluntary contributions. For example, in the United States, people who would otherwise have been more than willing to donate their genetic samples to promote scientific progress are now beginning to demand payment, once they realize that these corporations intend to profit by patenting their genetic materials.[133]

Free Markets and Public Science

For some of Milton Friedman's acolytes, public science is not at all objectionable—so long as it can be harnessed to corporate interests. Just consider the relationship between public science and the pharmaceutical industry. The government pays for extensive research to explore new avenues of medical research. Once public science identifies exciting new opportunities, private corporations can claim the fruits of this research as intellectual property.

The education of the former speaker of the House of Representatives, Newt Gingrich, symbolizes this belated awareness on the part of corporate-connected conservatives. Driven by their antigovernment ideology, Republicans under Speaker Gingrich proposed to set about slashing virtually every form of government expenditure when they first took over the House of Representatives in 1994. Toward this end, they proposed to cut funding for the National Institutes of Health (NIH) by 5 percent each year for five years. Fortunately, cooler heads prevailed.

The more pragmatic Republican Representative John Porter, chairman of the House Appropriations subcommittee that funds the NIH, was horrified. "This was insanity," he recalled. He quickly summoned scientific and medical experts to a session with the speaker, and something exceedingly rare occurred. Mr. Gingrich acknowledged he was wrong and the issue became how much to increase the NIH budget.[134] Well, not exactly. Rather than admitting an error, Mr. Gingrich brazenly turned around and took credit for nurturing the NIH. For example, in an op-ed piece in the *Washington Post,* the former speaker proclaimed:

> The highest investment priority in Washington should be to double the federal budget for scientific research. No other federal expenditure would create more jobs and wealth or do more to strengthen our world leadership, protect the environment and promote better health and education for all Americans. For the security of our future, we must make this investment now. . . .
>
> When I became speaker of the House in 1995 and we began to work toward a balanced budget, we were prepared to cut discretionary spending almost everywhere. At the request of Rep. John Porter, I met with the research vice presidents of all the major pharmaceutical and biotechnology firms. Even though they were overwhelmingly ideologically conservative, pro-free market and profit oriented, they unanimously agreed that the engine that was driving new medicines, new jobs and new profits was federal investment in basic scientific research and that American entrepreneurs would run out of new products and new services without that basic research.[135]

Of course, we have a peculiar interpretation of the meaning of pro-free market. According to the Gingrich school of thought, corporations should be free to market the public science that they appropriate under the government's vigorous protection of their presumptive intellectual property.

A few months before Gingrich published his op-ed piece, Gary Becker, a Nobel Prize-winning economist who spent much of his career at Milton Friedman's university, devoted his column in *Business Week* to the same subject.[136] It drew upon a then unpublished study by two University of Chicago economists, Kevin Murphy and Robert Topel, that estimated the value of the six-year increase in the expected lifetime of people in United States between the years 1970 and 1980.[137] Assuming a $5 million value on a human life—a relatively conventional estimate—they calculated that the value of the increase in longevity was equal to six times the 1999 gross domestic product of the United States. Of course, public medical science cannot take credit for all of this increase in longevity, but its contribution cannot be denied.

Because the expense of funding public science is so low compared to its benefits, this study concluded that such expenditures represent an extraordinarily productive investment. While the benefits of medical research are undeniable, the question remains as to how much of the profits from this research properly belongs to the corporations that patent this research. Some public officials have raised this question by modestly suggesting that the some of the profits from public research should return to the government.

When a private company gets an exclusive license to market a product resulting from federal research, the government is supposed to earn a royalty and the product price is supposed to be reasonable. In fact, the government has been shockingly negligent on both counts. Thus far, the government has done virtually nothing to recapture any of the benefits flowing out of research that it has supported.

Conveniently in their defense, the pharmaceutical companies invoke one of the strongest arguments against intellectual property; namely, that identifying the exact contribution of each individual party that had an impact on the final product is impossible.[138] The industry presumably interprets their claim to mean that since nobody could quantify the exact contribution of the government, the government should be content to take little or no royalty from the research.

Federal officials have not challenged the industry in this respect. In fact, they have not even bothered to keep track of the products, including drugs, that have profited from federally funded research. A 1995 study done at MIT found that of the 14 new drugs the industry identified as the most medically significant in the preceding 25 years, 11 had their roots in studies paid for by the government. In 1999, a preliminary report by the inspector general's office of the Department of Health and Human Services found that as many as 22 percent of discoveries financed by the federal health institutes were not reported by universities, as is required. More than 2,000 inventions developed with government money were reported to the health institutes last year, but officials told the *New York Times* that they had no idea which, if any, companies had licensed those inventions, or how they were being used.[139]

Even if the government could keep track of the products, determining a reasonable price is all but impossible without the help of the pharmaceutical companies themselves. According to a government report, "At present, the PHS [Public Health Service] has no established mechanism or standards for reviewing the reasonableness of prices for products marketed under exclusive licenses and lacks the legal authority to enforce its policy in cases where prices would be deemed unreasonable."[140] Federal agencies could only acquire the necessary information with the full cooperation of the company producing the drug. Not surprisingly, industry has stoutly resisted government efforts to get involved in pricing.[141]

Private Profit from Public Science

One particularly glaring case of abuse concerns the appropriation of public science in the creation of alglucerase, a medicine that Genzyme, Inc. markets under the brand name Ceredase for the treatment of Gaucher disease. This potentially fatal malady attacks between 2,100 and 11,000 people in the United States, mostly Ashkenazi Jews of Eastern European origin. Researchers at the NIH discovered the enzyme defect that causes the disease, which involves an inability to break down fats. They also devised a method for harvesting the enzyme from human placentae, for which they received a patent in 1975. Finally, they discovered a chemical modification that greatly improved the effectiveness of the enzyme. This modified form became alglucerase. Federal research indicated that a dose of 60 units per kilogram of body weight administered biweekly for a year resulted in rapid patient improvement, although patients would require indefinite treatments at a lesser dose.[142]

In addition to the research that took place within the government's own laboratories, the public invested almost $1 million in contracts with the New England Enzyme Center at the Tufts University Medical School to supply the government with sufficient quantities of the enzyme to continue the research process.

In 1981, Genzyme, then a new firm whose founders included researchers from the same New England Enzyme Center, took over the contract to supply the enzyme. The contracts with Genzyme over the next 11 years totaled nearly $9 million. This figure does not include the costs of decades of research by talented federal scientists or the work by researchers at universities or other private research institutions whose efforts culminated in alglucerase.

Ordinarily, the company could not have won the patent on the drug, because the federal government held the key patent. In 1983, Congress passed the Orphan Drug Act to give companies incentives to produce medicines to treat diseases that affected relatively few people. Genzyme quickly took advantage of this law and won a patent in 1985.

The Office of Technology Assessment estimated that Genzyme spent approximately $29.4 million on research and development for placental alglucerase over the decade prior to its approval for marketing in 1991. The company recouped $5 million from patients prior to approval of the drug, while it was still investigating the medicine's efficacy.

Once the drug won approval, Genzyme profited handsomely. Given the retail price of the medicine, treatment for the first year can cost between $71,160 and $552,760. A mere 60 patients paying the higher cost would more than repay all of Genzyme's initial outlays. Even at the lower cost, the company would only require 400 patients to recoup its development costs.

Of course, producing the drug entails some expense, but then again, these patients would have to continue purchasing lesser doses throughout their lifetimes, creating huge profits. By 1991, the company's alglucerase sales were already approaching $40 million; within less than a decade, sales exceeded $500 million. Building on this drug, the company became one of the five leading biotechnology companies in the world.

Although public science made this medicine possible, taxpayers pay for this research again each time the government picks up the tab for patients on Medicare who pay the inflated price for the drug. This case suggests why a conservative, such as Mr. Gingrich, is calling for more public research.

The Muted Call for Basic Research

Despite the widespread benefits to private industry from public research, neither Newt Gingrich nor Gary Becker are calling to reestablish the previous level of funding for basic science. Instead, they are recommending the sort of applied science that can be appropriated quickly by the corporate sector. Basic science—the sort of science that goes on without an eye to a particular application—remains neglected.

Basic research is less popular because of the long lag time between the initial scientific work and the final products. Recall the earlier estimates that most basic research requires 20 to 30 years before it blossoms in the form of final products. While applied research requires less time, it rarely leads to the revolutionary technological improvements that basic science makes possible.

I do not mean to imply that applied research is not useful. Of course, it is. The advances that applied research makes possible build upon earlier basic research. Although a balance between basic and applied research is important, the current emphasis on intellectual property along with the defunding of public research is creating a lopsided arrangement that is neglecting the basic research required for future technological progress.

Edwin Mansfield's surveys of the role of academic research in promoting new innovations provide some evidence regarding this reorientation of academic research results. In 1991, he published a study of the mean time interval between academic research and the first commercial introduction of related products or processes that result from that research.[143] In 1998, he updated his results. He reported that the lag was about seven years during the first period and six years in the second period.[144] Mansfield commented that

> this observed cut in the average lag may reflect a change in the nature of academic research. If the universities are doing more applied and short-term work, often geared toward relatively quick applications, the average lag would be expected to decline, but this would not necessarily mean that fundamental knowledge is being translated into products and processes more quickly.[145]

Mansfield's estimates do not contradict the earlier studies that estimated the lag between basic research and final products. Mansfield was asking a slightly different question. He was not studying the time lapse between basic research and the final product, but between academic research, which could be either basic or applied, and the final product. A researcher studying the contribution of research in Canadian universities to the development of technology concluded that much of what now counts as basic research is not basic at all, but merely "directed work in solving questions arising late in a product-development cycle."[146]

The call for more public science is also instructive for what it leaves out. While public science certainly will play a crucial role in the discovery of useful medicines, the intended purpose of this line of research is to advance private medicine—in which people receive individual treatment from a doctor. Despite the wonderful achievements of modern medicine, the real advances in life expectancy do not come from private medicine, but from advancements in public health.[147] A month-long strike by doctors in a densely-populated urban area would do far less harm to the residents than a comparable walkout by garbage collectors.

Like most forms of public investment, public health has suffered from terrible neglect in the United States. In the words of Laurie Garrett, author of a panoramic study of the decline in public health: "It took centuries to build a public health system, and less than two decades to bring it down. Once the envy of the world, America's public health infrastructure was, at the end of the twentieth century, indeed in a shambles."[148]

The full consequences of erosion of the public health system will not be felt until the nation faces an emergency, such as the rapid outbreak of a dangerous epidemic for which the system is not prepared. The anthrax scare of 2001 should have brought home this point.

Aside from the possibility of investing more in public health, had the society been free to divert public research from the effort to develop better medicines to discovering alternative methods of disease prevention by restricting the output of harmful products and emissions that the corporate sector creates, the health benefits could have been far more substantial. Instead, with the partial exception of tobacco, public medical science mostly directs its research toward finding medicines that treat the diseases once they have taken hold.

The difference between science directed toward preventing disease and science that aids in the creation of a pharmaceutical product is obvious. The latter science provides a profit for a corporation, while regulating the damage that corporations cause merely represents a nuisance from a business perspective.

The call for more public science neglects to mention still another cost of the present system of intellectual property rights. Because the profits from corporate science are skyrocketing for reasons that I have already discussed, industry can afford to hire top-flight scientists away from government jobs.[149] As a result, government agencies require more funds just to produce the same output.

The foregoing does not deny that public science will lead to significant advances in medicine. It certainly will, although the benefits from the new medicines will be limited by the problems created by intellectual property rights. More important, to the extent that the need to augment profits for private business constrains what scientists can and cannot do, the social and scientific benefits that flow from that science will be restricted.

Get the Lead Out

The history of tetraethyl lead, the poisonous gasoline additive that has since been banned, brings together a number of threads in this book, including the point I just mentioned about the rationality of preventing rather than curing illness. Intellectual property rights were a central factor in the initial development of this lethal product. The early research on gas additives actually favored alcohol, which could be made from agricultural waste products. The opportunity to gain a monopoly through patent rights was the main advantage of lead-based additives. According to an in-depth report by Jamie Kitman, a columnist and editor for *Automobile Magazine:*

> Farm alcohol was one thing, but a patentable process for production of petroleum-derived alcohol—a possible money-maker—was quite another, one of considerably greater interest to the corporation. . . . GM couldn't dictate an infrastructure that could supply ethanol in the volumes that might be required. Equally troubling, any idiot with a still could make it at home, and in those days, many did. And ethanol, unlike TEL [tetraethyl lead], couldn't be patented; it offered no profits for GM. Moreover, the oil companies hated it, a powerful disincentive for the fledgling GM, which was loath to jeopardize relations with these mighty power brokers. Surely the du Pont family's growing interest in oil and oil fields, as it branched out from its gunpowder roots into the oil-dependent chemical business, weighed on many GM directors' minds.
>
> As the management expert P. F. Drucker described it many years later, "GM, in effect, made money on almost every gallon of gasoline sold, by anyone."[150]

The use of lead as a gasoline additive also illustrates the threat to the integrity of science under corporate control. Kitman's article documents the tragic reality of the science surrounding lead. A 1985 EPA study estimated that as many as 5,000 Americans died annually from lead-related heart disease prior to the country's lead phaseout.[151] The corporations fought tooth and nail against those who would question the safety of lead. They spread their largess to compliant institutions whose researchers would confirm the safety of lead. In addition, the corporations vilified those researchers with the temerity to raise questions about the public health risks of lead. Even though many in the scientific community understood that the use of lead represented a significant threat to human health, government agencies bowing to corporate pressure found time and time again that the spread of lead through burning fuel was benign, undermining the work of legitimate scientists.[152]

Finally, the sad history of lead offers another demonstration of the lock-in phenomenon. Once the infrastructure for adding lead to gasoline was fixed in place, the creation of alternatives became very difficult. Too much was invested in physical infrastructure, too many people had vested interests in continuing with the current situation, to easily reverse the initial decision to base the automotive fuel system on lead additives.

In the case of pharmaceuticals, concerns about the protection of intellectual property prevented the industry from saving lives. In the case of lead, concerns with intellectual property set off a chain of events that directly caused death. Only after the public began to react to the overwhelming evidence of the dangers of lead did the government take any action.

CHAPTER FIVE

ECONOMICS VERSUS THE NEW ECONOMY

A Brief Introduction

Throughout this book, I have mentioned ways that intellectual property corrodes the scientific process. Since virtually every economist will agree that scientific progress is a key to advancing economic productivity, anything that prevents science and technology from progressing impedes the economy. In this chapter, I will go into more detail regarding these and other pathways through which intellectual property harms the economy. I will also explore some of the misperceptions that obscure the vision of many people who try to understand the nature of the new regime of intellectual property.

First and foremost, the tightening grip of intellectual property rights widens the gap between rich and poor, making the rich richer and the poor poorer. To the extent that the government strengthens the hold of intellectual property rights even further, still more wealth and income will shift away from the rest of society to the powerful few that now control the bulk of intellectual property.

This trend toward greater inequality creates a further problem. To begin with, more and more, economists are coming to understand how a highly unequal distribution of income and wealth hampers economic growth.[1] When the less fortunate people in society become too downtrodden, they become less able to contribute to the productive process. To make matters worse, extreme disparities in wealth require more resources for controlling people, through police and prisons, as well as private expenditures for security. Such expenses represent an enormous drain on society. For example, the State of California presently is spending more public money on its prison system than on higher education.

Intellectual property rights are incompatible with the forces of competition, which are supposed to be the lifeblood of a market economy, for another reason. Most economists would agree that although scientific knowledge determines the maximum possible performance of a market economy, for a society to achieve anything near that level requires businesses to be run efficiently. They would also concur that forces of competition push businesses to adopt the appropriate management techniques. This problem explains why so many believers in the market opposed patents, especially in the nineteenth century.

I would add that excessive competition, such as occurs during a severe depression, creates chaotic conditions that also interfere with efficiency. At the risk of a gross simplification, the economy of the United States is in the process of dividing up into two unequal parts. While one part of the economy enjoys a highly profitable existence under the protection of intellectual property, the rest of the economy tends to suffer under a withering hypercompetition that makes an excessive number of firms limp along at the brink of extinction. Neither the monopolistic protection of intellectual property rights nor the heat of hypercompetition is conducive to a well-working economy.[2]

Finally, I will discuss how economic theory proves that markets are incompatible with the purchase and sale of information—the very sort of markets that are central to a good deal of intellectual property rights. Recall Robert Merges's assertion that intellectual property was "born as a response to market failure."

The Economics of Public Goods

I want to return to Merges's observation about market failure. The central proposition of conventional economics is the idea that markets are efficient because competition drives prices down to the marginal cost—the cost of producing one more unit of a good. According to this logic, by setting a price on a good that excludes all but those who can afford that good, markets ration goods efficiently.

Much of economics purports to show that allowing a competitive market to set prices for commodities is the most efficient method to distribute goods and services. This rationale is not wholly convincing to all economists because of the widespread effects that prices fail to reflect. Pollution and the depletion of resources are familiar examples of this sort of deficiency of the price system. Others object to this system on the grounds of equity.

Putting such concerns aside, most economists admit one important exception, when the logic of markets does not make sense: rationing by price creates inefficiencies in all cases in which the marginal cost is very low while the unavoidable fixed cost, which a firm must incur regardless of the quantity of goods that is sold, is high. Thus, according to conventional economic theory, goods with zero marginal costs should be treated as public goods, meaning that they should be given away without cost rather than be sold as commodities.

When a person consumes a typical good, less remains for the rest of society. In the jargon of economics, goods that have zero marginal costs are called non-rivalrous. This strange-sounding term is meant to convey a simple but crucial point. If I eat a can of soup, less soup remains for you. Soup, then, is rivalrous.

Information is the ultimate non-rivalrous public good. It is not scarce in the same sense as other commodities are. After all, people do not use up information in the same way that they consume food or fuels. You do not give up your possession of information when you give it to another person. Information is not depletable. If I consume a bit of information, I do not impair your access to information. With information, you *can* have your cake and eat it too.

The price system does not work efficiently even when the good has a small but nonzero marginal cost if the fixed costs are large by comparison. I refer to such goods as quasi-public goods. In the case of public goods, the best arrangement is to have the government pay for their production with what are called lump-sum taxes—taxes that are imposed on people regardless of whether they use the commodity or not.

Information is a classic example of a public good. If I learn algebra or Chinese, I do not reduce the amount of algebra or Chinese for the rest of the world to learn. The logic of public goods demonstrates that information, including science, should be provided by society as a whole. Those accustomed to accepting the logic and the discipline of the market without question might well ask, Why should the government pay to produce a good that it then gives away? Shouldn't the government at least charge a user fee?

No. Economics teaches that goods are supposed to be priced at their marginal cost. If the state collected a tax or user fee each time a person used a unit of a public good, the tax or fee would be inefficient. The payment of a tax or user fee would discourage the use of the public good, even though such usage would impose no cost on the rest of society. The resulting wasted opportunity would create an inefficiency, according to the logic of economics.

Imagine a government that had a reservoir of information. Remember that markets are supposed to work efficiently because they encourage people to economize on scarce resources, such as food or fuel. Such economizing, in turn, is supposedly beneficial because it induces people to conserve scarce resources.

But what benefit could the rationing of information serve? After all, the pool of information is not subject to the logic of scarcity. Once the government possesses that information, distributing it would not cost anything. As a result, the government could provide a benefit to people at no cost whatsoever.

The role of efficiency within the study of economics deserves further comment. The economic rationale of laissez faire rests on the idea that the market can ration goods efficiently. When faced with a proposal to intervene in market outcomes, proponents of laissez faire counter that any such action would create inefficiency.

The case for public goods also rests on the logic of efficiency. While people can use economic theory to oppose interference with the market for private goods on the grounds of efficiency, they should also accept the logic of economics to support the public provision of public goods, unless they are just being hypocritical and only invoke considerations of efficiency for purely ideological reasons.

Of course, the initial production of public goods consumes resources. Society will still have to decide which public goods it will produce, but the market is ill suited to such a task. Take the example of a movie—admittedly a good with a typically low informational content. The production costs alone of a film may run from $10 to $200 million, but the cost of showing you the film—the marginal cost of admitting you to an already scheduled screening of the film—is virtually zero if some of the seats are empty. Your attendance may create some nominal costs, perhaps by spilling your popcorn, but you will certainly not create costs anywhere near the typical price of a ticket to the movie.

Keep in mind that shifting informational commodities to a public goods status may actually save money. For many informational goods—say a long-distance phone call—the greatest cost of the transaction is not the call, which actually costs the phone company virtually nothing. The cost associated with billing customers for the call is far greater than the call itself.[3] Even in the movie theater, the selling and then tearing of the tickets probably represents one of the larger costs imposed by your attendance.

Of course, merely declaring informational goods to be public goods does not solve all economic problems. Society will still have to decide how much of its resources will be devoted to opening new theaters, making new films, or adding new long-distance capacity. The public will somehow have to develop new institutions appropriate for making such decisions. Perhaps most of all, society will have to raise the level of education so that the public will be prepared to make such decisions in an intelligent manner.

The fact remains that the market is an inefficient mechanism for handling the sort of commodities that qualify as public goods or even quasi-public goods. Intellectual property rights are capable of allowing such goods to market at a price considerably above marginal costs, but they do so only by distorting the economy. In short, the situation represents a classic example of a market failure.

Public Goods and the Inappropriateness of Economic Theory

Around 1995, after decades of very low productivity growth, productivity in the U.S. economy dramatically increased. Immediately, a number of economists and business writers seized on the idea that we were witnessing the emergence of a New Economy. At last, the computer revolution, after years of astounding technological wizardry without any visible effect on the overall economy, was suddenly transforming the economy. They proclaimed the beginning of a new era that transcended the limitations of the economy as it was previously known.

Popular writers also sensed a fundamental change in the economy. Impressed that production costs are virtually nil in much of the economy, they believed that they were witnessing a technological triumph in which information suddenly had become the center of the economy. They breathlessly referred to a weightless economy.[4]

Tom Peters, the management guru, derided old-line businesses as "Lumpy-object purveyors."[5] Even Alan Greenspan, chairman of the Board of Governors of the Federal Reserve System, is fond of rhapsodizing about how modern production techniques are making the economy lighter and lighter:

> The world of 1948 was vastly different from the world of 1996. The American economy, more then than now, was viewed as the ultimate in technology and productivity in virtually all fields of economic endeavor. The quintessential model of industrial might in those days was the array of vast, smoke-encased integrated steel mills in the Pittsburgh district and on the shores of Lake Michigan. Output was things, big physical things.
>
> Virtually unimaginable a half century ago was the extent to which concepts and ideas would substitute for physical resources and human brawn in the production of goods and services. In 1948 radios were still being powered by vacuum tubes. Today, transistors deliver far higher quality with a mere fraction of the bulk. Fiber-optics has [sic] replaced huge tonnages of copper wire, and advances in architectural and engineering design have made possible the construction of buildings with much greater floor space but significantly less physical material than the buildings erected just after World War II. Accordingly, while the weight of current economic output is probably only modestly higher than it was a half century ago, value added, adjusted for price change, has risen well over threefold.[6]

The rhetoric of weightlessness has a strong political undercurrent, emphasizing the economic contribution of entrepreneurial and professional people at the expense of those who do more manual labor. Recall Ayn Rand's claim that "Patents and copyrights are the legal implementation of the base of all property rights: a man's right to the product of his mind."[7] Rand warned that collectivists were in the process of attacking intellectual property rights in an effort to undermine capitalism.[8] Interestingly enough, Greenspan was an ardent admirer of Rand. She apparently reciprocated. Her defense of intellectual property appeared in a book that mainly contained articles that she had published earlier, except for a couple of essays by none other than Alan Greenspan. Rand, however, never made the leap to the idea of weightlessness.

The initial prophet of weightlessness was George Gilder. More than a decade earlier than Greenspan's pronouncement, flush with the victories of the Reagan revolution, Gilder extravagantly proposed: "The central event of the twentieth century is the overthrow of matter. In technology, economics, and the politics of nations, wealth in the form of physical resources is steadily declining in value and significance. The powers of mind are everywhere ascendant over the brute force of things."[9]

For Gilder, weightlessness served as a decisive refutation of Karl Marx's emphasis on labor. In Gilder's view, ideas—what now more commonly would fall within the purview of intellectual property—rather than the efforts of workers create wealth and value. This devaluation of blue-collar labor resonated in the fulsome praise heaped upon the rise of the New Economy, supposedly driven by ideas and information. Greenspan's description of weightlessness was more measured and less blatant, but not qualitatively different from Gilder's.

Do not be misled by the appealing rhetoric of the weightless economy, which is not nearly as immaterial as it might seem. *Forbes* published an admittedly exaggerated article that made the claim that to create, package, store, and move two megabytes of data requires one pound of coal.[10] The authors of this article, Peter Huber and Mark Mills, estimated that more than 13 percent of power output in the United States was being used to manufacture and run computers and the sprawling information technology infrastructure. Later, they claimed that that estimate was too conservative.[11]

Interestingly, Huber and Mills work closely with Gilder. George Gilder's Gilder Technology Group publishes the Huber-Mills Digital Power Report newsletter. The newsletter's Web site contains Gilder's fulsome praise for the authors.

A study done at Lawrence Berkeley National Laboratories found that the estimate of Huber and Mills is wildly excessive.[12] Rather than representing an objective study, it seems to have been intended to give comfort to the energy industry. In his presidential campaign, George W. Bush gave a speech in Saginaw, Michigan, using the estimate of Huber and Mills to support his call for an expansion of oil drilling in the Arctic National Wildlife Refuge.[13] Not surprisingly, the coal industry funded much of their work.

Despite its limitations, the estimate of Huber and Mills does serve as a reminder that the resource costs of the supposedly weightless economy can be substantial. For example, the weightless economy depends on colossal server farms. One, being constructed near Seattle, covers about 13 acres.[14] These server farms have power concentrations of 100 watts per square foot. Ten square feet consume enough power to supply a typical home. These data centers are expected to cover an estimated 50 million square feet by 2005, but even so their demand will amount to slightly more than 1 percent of the U.S. electricity demand by that time.[15]

The material demands of the New Economy go well beyond fossil fuels. The United States used about 1 billion metric tons of materials in 1990, such as iron, copper, sulfur, and phosphorus, and hydrocarbon fossil fuels, as well as other materials that are mined and used in the production of goods, but excluding some (crushed) stone that is used to build roads and other structures.[16] Despite the growth of the New Economy, the resource requirements of the American economy have no doubt grown since then.

Nor should the other enormous environmental costs of the production of high technology be forgotten. Production of silicon wafers may not require the huge smokestacks associated with traditional rustbelt industry, but it relies on large infusions of toxic solvents that have poisoned much of the ground water of Silicon Valley.[17] The plastic alone in a typical computer requires 1 ½ gallons of crude oil and 300 cubic feet of natural gas to manufacture, according to calculations made by the Technical University of Denmark.[18]

Finally, although low-wage workers may not command much wealth and income, society is very dependent on their contribution. Despite all the talk about the weightless economy, the food that these pundits eat and the clothes that they wear probably come from low-wage labor. Rather than fantasizing about a weightless economy, they would do better to pay closer attention to the invisible economy of the farm workers, the sweatshop workers, and maybe even the nannies in their homes—the invisible workers who live from hand to mouth without the luxury of fantasizing about a weightless economy.

Greenspan's remarks are even more interesting for what they don't say. He never mentions the concept of public goods or the possibility that rationing goods with low marginal costs can lead to economic inefficiency. Instead, Greenspan's rhetoric appeals to the rapid run-up in the stock market at the time and the incredible fortunes created by new start-up companies—more often than not built around a foundation of intellectual property rights.

Greenspan seems to attribute this wealth to the ingenuity and creativity of a small group of entrepreneurs and technologists, without acknowledging that the public sector had laid the groundwork for the New Economy during the period in which government more generously funded university research. Even the Internet, as well as much of the computer industry, owed an enormous debt to the Defense Department, which funded virtually all of its basic technologies.

Greenspan also seems to suggest that the upsurge in wealth permeates the entire society, forgetting that a considerable amount of high technology utilizes very low-wage workers. Finally, as I suggested above, Greenspan never acknowledges that a good deal of the wealth of the new technology reflects the power of intellectual property rights.

Nowhere do we hear him utter words such as Microsoft, Nike, or Monsanto. Yet, what makes the value of the products of these giants soar relative to the weight of what they market is the power of their intellectual property rights. My reading of the situation is that the weightlessness of the New Economy is a product of the weight of intellectual property rights, which bear down heavily on the rest of society.

In one sense the rhetoric of weightlessness makes a great deal of sense. What drove the economy of the late 1990s was the stock market, which defied the logic of the material world. Driven by wild dreams of instant wealth, the stock market began a climb into the airy reaches of fantasy.

For a while, the real economic world seemed to adapt to the fantasy of the stock market. Wages and salaries for some of those who were associated with the dot.com phenomenon became as unrealistic as the stock market. This distortion, combined with the power of intellectual property rights, created a highly distorted economy.

This process began even before the 1990s. For example, between 1979 and 1989, the top 1 percent of households got 70 percent of all of the increase in income.[19] Rather than celebrating the achievements of the weightless economy, I believe that we would be better served by raising concerns about the burden that the weightless economy poses for the rest of society, which did not share in the prosperity.

On Weight and Value

The rhetoric about a weightless economy conflates issues of technology, value, market power, and productivity. For example, a pair of tennis shoes may increase in value relative to its weight either because Nike has developed new technology that allows the same shoes to have more valuable properties or because the shoes have become lighter. Alternatively, and probably more realistically, Nike could sell the very same shoes at a higher price just because of sheer market power. All of these possibilities would raise the market value of a pound of shoe. Aggregate data does not allow us to distinguish between these different possibilities.

Of course, weight as such should have little direct relevance to value, except in particular situations. For example, miniaturization creates additional costs, as in the case of a notebook compared to a desktop computer. However, other examples of a rising value per pound are totally unrelated to the complexity of production. For example, in the case of tennis shoes, prices have been soaring over the decades, even though the actual production cost is only a tiny fraction of the sales price.

Almost two decades ago, David Slawson, pointed to a related example in a book entitled *The New Inflation: The Collapse of Free Markets.* He observed that companies that sold candy bars managed to raise prices from a nickel in the late 1950s to 35 cents in the early 1980s through a judicious manipulation of the size of the candy bar as well as the packaging that contained it. At first, the price of the candy bar would remain fixed while its size diminished. Later, the company would increase the size of the candy bar along with the price. Then, the company would gradually reduce the size again. Frequently, the company would mask the reduction in size by keeping the dimensions of the packaging constant.[20]

So while Chairman Greenspan views the increasing ratio of prices to weight as an indication of the success of markets, Slawson condemned the same phenomenon as an exercise of raw market power. In effect then, the increasing cost per pound can reflect either some technological mastery that has allowed producers to put a greater value in the same mass of product or, in the absence of some technological change, sheer market power has allowed producers to charge more for the same product.

Traditionally, producers of manufactured goods managed to maintain high markups via accumulating market power. They blunted the effect of competition through strategies that many considered to be anticompetitive. Corporate consolidation was undoubtedly the preferred tactic at the turn of the century. For example, during the great merger wave, the *New York Daily Tribune* reflected the optimism that the mass of corporate consolidations had generated at the time (1901): "[A] new era has come, the era of 'community of interest,' whereby it is hoped to avoid ruinous price cutting and to avert the destruction which has in the past, when business depression occurred, overtaken so many of the competing concerns in every branch of industry."[21]

As the giant corporations gained control over industry, they transferred wealth from the public to themselves. Eventually the public rebelled. Congress passed the Clayton Antitrust Act of 1914 in response to public pressure. In addition, the government forced the breakup of a few giant corporations. At best, these actions put a slight dent in the rise of corporate power.

In general, government policy with respect to antitrust has fluctuated between acquiescence and modest resistance. As Dennis Mueller, a noted student of antitrust, has noted:

> The United States has the longest history of antitrust enforcement in the world, over 100 years now. Over this span, U.S. antitrust policy has waxed and waned in its vigor but on average the U.S. has had the toughest and most vigorously enforced antitrust statutes in the world. In recent years, the wisdom of this policy has been questioned, the government has taken a noticeably more circumscribed approach to enforcing the antitrust laws, and the courts have become more lenient in interpreting them.[22]

During the great merger wave at the turn of the century, when the major manufacturing industries of the United States consolidated, the value per pound of production certainly increased. I doubt that anybody at the time looked at the steel industry and marveled at how weightless it had become.

The present thinking is that the economic future no longer rests with high-weight, low-value products, such as steel or grain. Lacking market power, such industries enjoy modest profits at best, while they find themselves beset with intense competition. If, however, the steel industry won relief from competition as effective as that which intellectual property rights provide for computer software or Nike shoes, the pundits would be raving about the enormous economic success of the steel makers. Legislators would fall over each other in a rush to confer tax breaks and other benefits to help this vital sector of the economy enhance its growing prosperity. The industry, flush with funds, would return the favors by spreading its largess in the form of campaign contributions or advertising campaigns.

Today, commentators are less inclined to think in terms of market power than earlier generations did. Instead, they use the fashionable jargon of weightlessness to turn people's attention from such mundane matters. According to the rhetoric of weightlessness, huge profits flow to those who can best harness the power of the intellect. For those enraptured by a vision of a New Economy, what market power exists arises solely from the creativity and efficiency of corporations.

In contrast, this book has taken the position that the accumulation of market power in the form of intellectual property rights has facilitated an unprecedented transfer of wealth and income to a small stratum of society and that intellectual property is a major source of that market power. Theoretically, the antitrust laws should keep market power in check when competition fails to do so. Unfortunately, rather than maintaining even a pretense of a meaningful antitrust policy, government agencies are now tripping over themselves to strengthen intellectual property rights, while watching the first wave of worldwide corporate consolidations.

During the great merger wave at the beginning of the twentieth century, corporations organized among themselves to limit competition so that they could boost their markups. Although corporate consolidation is helping to increase markups in some traditional commodities, such as petroleum, intellectual property is probably the most important vehicle for increasing the gap between sales price and the cost of production in the present economy.

For example, between 1977 and 1996, the core copyright industries grew nearly three times as fast as the annual rate of the economy as a whole—4.6 percent vs. 1.6 percent. Some of this growth reflects the increasing output of these products, but some also results from the growing markups that the purveyors of these products enjoy. No wonder the value of output per pound is growing!

At the same time, the producers of basic commodities that sell to highly competitive markets have seen their profit rates erode. Even further down the economic ladder, real wages have fallen for those workers who perform the menial tasks required to keep the economy going. The homeless workers living amidst the wealth and luxury of Silicon Valley, whom I will discuss later, are a particularly poignant example.

In contrast, the purveyors of intellectual property find the government willing to go to great lengths to promote their prosperity by insulating them from competition. Fostered by the absence of any meaningful antitrust actions by the government, combined with the growing stranglehold of intellectual property rights, they accumulate unimaginable wealth. This process, however, is inflicting great damage on the social and economic fabric.

A Brief Note on Economic Theory

Alert readers have probably noticed an ambivalence about economic theory in this book. At times, I refer to what economic theory concludes about a particular situation. At other times, I dismiss what economic theory can contribute to a discussion.

In order to make myself clear, let me take a moment to explain my interpretation of the strengths and shortcomings of economic theory in relation to intellectual property. Economists have an excellent grasp of the nature of the simple transaction. For example, a person wanders up to a fruit stand and ponders whether to buy some grapes, an apple, or an orange. In a sense, the economist has a more formal way of indicating that the person will buy the fruit that will give the most benefit per dollar.

Many transactions are not that simple. They involve complex considerations. For example, the production of a fruit might involve the use of toxic chemicals that threaten the local water supply. A prospective buyer, knowledgeable about this problem, might refrain from buying the fruit in question, even though it might be preferable in other respects. Alternatively, the prospective buyer might have a personal relationship with the farmer that produces one of the fruits. As a result, non-economic factors will enter into the decision about which fruit to purchase.

The analysis of public goods follows directly from the economic logic applied to a transaction for a particular good. When the discussion shifts to the production of information, the transaction becomes very complex. In the production of apples, farmers adjust their production according to their best guesses about the probable supply and demand for apples. Years of repeated experience give farmers a degree of expertise that may be sufficient for them to make a moderately good estimate about the demand for the coming years.

In contrast, as far as the economics of decision making is concerned, the production of science is hopelessly complex. The apple farmer has a fairly good idea about how people will use the apples. The apple farmer has some idea about the size of the harvest—or at least an idea about how likely a very big harvest or a very little harvest will be. Scientists have no such information. If they are successful and their work bears fruit, other people are likely to use the knowledge that they develop in ways that they never anticipated. In short, science is a network process rather than an individual transaction. For this reason, conventional economic theory has relatively little to contribute in understanding how science develops.

Lacking theoretical guidance, economists become more prone to fall prey to exaggerated metaphorical explanations of the economy, especially because their training imbues them with an undue reverence for market processes. Their acquiescence lends credence to the notion of a weightless economy.

Information and Economic Metaphors

Despite all the loose talk of an information economy, most economists have failed to recognize how much the ground has been changing under their feet. In addressing the question of intellectual property, economic theory suffers from the same sort of confusions that are implicit in the misleading metaphor of the information superhighway, which suggests that information is trucked about like so much soap or canned soup. This approach seemed consistent with economists' well-established practice of restricting their conception of scarce resources to three seemingly well-understood entities: land, labor, and capital.

Economists presumed that markets could somehow measure each of these resources in an unambiguous way to ensure that each was put to the most effective possible use. The simplicity of this approach became less meaningful with each passing day as the conventional conception of scarce resources became increasingly nebulous.

For example, economists once regarded labor as nothing more than a set of interchangeable tools. In part, this approach reflected a commonly accepted outlook—at least among those who did not have to work for a wage—that labor was nothing more than an almost mechanical performance of predefined tasks. Our common language crystallized this perception, casually referring to workers as hired hands.

Happily, most economists have abandoned this dehumanized perspective, realizing that workers are heterogeneous. This evolution reflected the changing circumstances of economists themselves. In the early days of economics, no major economist, with the possible exception of Thomas Robert Malthus, famously associated with the idea of overpopulation, worked for a wage. Over time, as more economists became dependent on salaries, economists no longer believed that all wage workers were merely brute instruments of labor.

In order to come to grips with this expanded vision of the labor force, economists devised a new concept. Specifically, they invented a new factor, which they called "human capital," a theoretical quantity that is supposed to reflect the effect of the education and experience of a worker. Thus, human capital is separate from and in addition to the conception of the worker as a basic mechanical device.

You may find the idea of human capital to be a bit weird. I do. Certainly, the language is wonderfully ambiguous, mixing the idea of "human" with capital—an obviously inhuman concept. According to the imagery of human capital, we have a mix of the "human" aspect of labor—which, ironically, is comparable to the earlier inhuman vision of labor as a pair of hands or arms—together with an inhuman or capital part—reflecting education and experience, aspects of life that we normally associate with a humanizing influence.

Does the human being somehow give life to the capital? Or perhaps we should say that the concept of human capital dehumanizes humans to the level of capital. In order to be fully human, a person must enjoy ownership of a significant quantity of this particular form of capital.

The rest of humanity, lacking sufficient human capital, will supposedly function best as unthinking machines. For example, according to a statement from a 1995 round table of the Fraser Institute, a conservative Canadian think tank: "The most important job skill anyone must learn is 'not being insubordinate.'"[23]

Despite the metaphorical confusion, the theory of human capital does reflect an undeniable truth. Workers' education or experience can often amplify their productivity, even though formal educational qualifications frequently have nothing to do with what workers are expected to do on the job.

What about information? The "information" in the notion of an information economy is different from human capital. You might think of human capital as a particular kind of private information. The metaphor of an information economy mixes together all sorts of information: public information, private information, commercially available information, etc.

Our language, it seems, is not yet adequate to address the concept of an information economy. While most economists accept that information is both widespread and intangible, they tend to speak of information in metaphorical terms. For example, you may have heard people suggest that genetic information lies at the heart of the simplest biological life forms.

While some of the estimates of the contribution of information are excessive, information undeniably plays an increasingly important role in the economy. Unfortunately, economists are ill prepared to deal with the reality of an information economy.

Why Information Should Not Be Treated as a Commodity

Throughout this book I have pointed out numerous drawbacks in relying on the corporate sector to create what the law now treats as intellectual property. Disregarding for the moment the sort of intellectual property that is purely commercial—such as brand names and the like—I will refer to intellectual property as information out of deference to those who believe that information is the central asset in the present economy, as opposed to all previous economies.

Despite the confident talk about the information economy, I would not be far off the mark to assert that both information and science fall outside of the scope of conventional economics. Some background will be helpful in clarifying this assertion. Economic theory began with observations of conditions that resembled a village economy dominated by handicrafts. The famous Scottish economist Adam Smith, writing in the midst of the Industrial Revolution in the eighteenth century, mentioned nothing more modern than some craftsmen making pins.

The second component of early economic theory concentrated on understanding the impact of merchants trading with one another, using the tools of accountants. Although accounting theory was appropriate for merchants who bought and sold goods with a relatively quick turnover, the accountants never developed an adequate method of handling depreciation of fixed capital; that is, capital goods that depreciate over an extended period of time. As a result, conventional economic theory fails when it comes up against an economy with massive fixed costs and trivial marginal costs, meaning the cost of producing one more unit of the good. I have gone into detail about this defect in economic theory elsewhere.[24]

In effect, then, economics is the theory of capitalism without capital. Few economists took note of these gaping holes in the core of economic theory. This blindness was understandable. After all, the market worked well enough to satisfy most well-off people, except during periods such as the Great Depression. I attribute this good fortune to a combination of government intervention and business collusion, together with some early customs and traditions. During the twentieth century, when economics became a formal profession, these factors kept the economic system from falling into deep depressions, with one glaring exception.[25] Although the Great Depression profoundly shook confidence in the market, once prosperity returned, the profession quickly regained its confidence in markets.

If conventional economics is ill suited to handling an economy characterized by large investments in capital goods, it is even less adequate for dealing with the so-called information economy. As I have mentioned before, market relationships are particularly inappropriate for handling information. After all, the supposed purpose of property rights is to induce people to economize on scarce resources. Accordingly, most modern economists see their discipline as the science of allocating scarce resources.[26] As a result, property rights make no sense in the absence of scarcity.[27]

But information, a major constituent of intellectual property rights, is not scarce. As Kenneth Arrow recently noted, "Patents and copyrights are social innovations designed to create artificial scarcities where none exist naturally."[28] In spite of the efforts to make information artificially scarce, economists realize that information differs from scarce goods, such as detergents or canned soups.

These scarcities, however, serve no social purpose whatsoever. In fact, using the market to exclude people from access to information is self-defeating. It does not increase the supply of information. It only spreads ignorance. Nor does my consumption of information detract from the access of anybody else; it may even add to the pool of social information, possibly creating an advantage for others. As a result, fields of research are very different from agricultural fields. While exclusivity is imperative in the farmer's field, it makes no sense whatsoever in science. After all, the more information that I gather, the more potential information is available to you.

For example, if you let me read your book or use your computer program, you may benefit from sharing the fruits of my experience. In fact, unlike so-called rivalrous goods, which can be used up, the more that people partake of the supply of information, the greater the total stock of information becomes. In short, using information can spawn more and better information. For instance, as a scientist learns more about her field, she has more to share with others. While scientists might compete with each other for the priority of a finding, the discovery of one enriches all.

While Adam Smith may have been silent about the modern technologies of his own time, he made an important observation that bears on the modern information economy: "The navigation of the Danube is of very little use to the different states of Bavaria, Austria, and Hungary, in comparison of what it would be if any of them possessed the whole of its course till it falls into the Black Sea."[29] What he said about navigation also holds for information. Openness, without the restrictions of property rights, makes information all the more productive.

A decade ago, I coined the term "metapublic goods" to emphasize that the case for keeping information beyond the reach of the market was even stronger than that for the typical public good.[30] A pure public good has a zero marginal cost, meaning that the consumption of the good by an individual imposes no cost on the rest of society. Unlike most physical goods, the use of information and ideas goes beyond this neutral cost, because it actually enriches the community.

I cannot emphasize this point enough: The concept of scarcity is absolutely irrelevant to information. The more the law restricts people's access to information, the less information will be available.

As a result, economics, which economists themselves define as the allocation of scarce resources, has little to offer in an information economy. Information is not scarce, except to the degree that society allows agents to create artificial scarcity through secrecy and property rights. More to the point, as the economy becomes increasingly dependent on information, the traditional system of property rights applied to information becomes a costly fetter on our development.

Of course, most conventional economists may take issue with this perspective; however, conventional economics is seriously deficient when it comes to the analysis of information. In the words of Richard Nelson:

> Economists clearly are schizophrenic in their treatment of property rights in technological knowledge. In standard microeconomic theory, technological

> knowledge is assumed available to all. Most of the growth models . . . implicitly treat technological knowledge as public. On the other hand, in their analysis of the effect of the patent system, or of technological rivalry among firms, economists treat (at least some aspects of) technological knowledge as proprietary.[31]

In short, to the extent that economics supports the treatment of information as private property, the discipline should modify its understanding of its purpose. No longer can economists pretend that they are merely analyzing the allocation of scarce resources; in addition, they are advocating the creation of scarcity.

The Strange Economics of Information

Although I contended that information and science fall outside the scope of conventional economics, the study of economics does provide some guidance about the nature of markets for information. Recall that markets are supposed to work by equating price with marginal cost.

According to the logic of economics, the market is supposed to work because it sets a price on a good that excludes all but those who can afford to get the most use from that good. But what purpose does this rationing serve for non-rivalrous goods? Excluding people from information does no good whatsoever.

Kenneth Arrow observed that in the case of information, even though the original cost of gathering the information may have been substantial, the cost of transmitting this information on to others is minimal.[32] In other words, the marginal cost of information is effectively zero. For example, the discovery of an important natural law may be a heroic event, yet the cost of sharing it with the rest of the world is insignificant. A scientist could simply post it on the Internet for all to read.

Arrow pointed to still another flaw inherent in the market for information that further undermines the case for a market in information. Markets need informed customers. In this sense, information is different from other goods. In shopping for clothing, for example, we can get information about the clothing by browsing through the store. We can even try on an article of clothing to see how it looks or feels. In the case of information, by contrast, the information about the product and the product itself are identical. We can possess an informational product merely by learning about it. Consequently, the owner strives to keep the information as secret as possible, preventing potential customers from shopping for information in an informed manner.[33] As a result, an informed market in information is a contradiction in terms.

Arrow's logic is indisputable. To begin with, since the marginal cost of information is zero, information should be free. Besides, even if we want to work within a market system, the market cannot work well because people cannot know the value of the information that they contemplate buying without first acquiring that information.

As a result, our basic concepts of economics are of no use in an information economy, except in one paradoxical sense. Even though economic theory is severely biased toward markets, according to the criteria of economics, information should not be treated as private property. As a result, when information becomes the dominant resource, the laws of economics tell us that the laws of economics themselves are invalid, since information should not be priced as a scarce resource.

So we can see why Robert Merges concluded earlier that "patents (and perhaps intellectual property generally) were born as a response to market failure."[34] Unregulated markets for information just will not work for the reasons that we have just discussed. Markets for information are bound to fail without some sort of special protection.

So here is the crux of the problem of information. According to basic economic theory, in a free market the price of information will approach zero. According to the basic theory of economics, inventors should ideally earn an income high enough to induce them to invent until the marginal cost of their inventive efforts equals the marginal value of their invention.

Measuring the marginal value of an invention is, of course, impossible, especially since an invention devised for one purpose will often create more value somewhere else. Consider the ubiquitous transistor. AT&T may well have more than recovered its investment in the transistor, but that savings is minute compared to the impact of the transistor on the rest of the economy. The company's investment was profitable only because AT&T was so large. If a small firm had invented the transistor, it would not have been likely to be able to enjoy enough saving to make such a massive research effort profitable.

In effect, then, a free-market solution is impossible. On the one hand, if information is priced appropriately, those who create the information will be underpaid, if they are paid at all. As a result, too little effort will be put into the creation of information. On the other hand, if those who create information earn the appropriate amount, information not will be priced appropriately, leading to an underutilization of information.

The patent system opts for the underutilization of information in the hopes that the greater output of information will be best for society. Even with the extraordinary rewards that holders of intellectual property enjoy, economists estimate that the optimal rate of investment in research and development in the United States would be four times higher than what that economy now experiences.[35]

Could this estimate mean that intellectual property rights need to be even stronger? I would dismiss that possibility. I would suspect that a stronger system of intellectual property rights might be just as likely to discourage additional investment in research and development because of the increasing tendency to litigate over who is entitled to the rewards.

I would favor a second interpretation. Because of the complications imposed by intellectual property, at any moment the economy is underutilizing existing information. As a result, the data suggest that you need several times more research and development just to accomplish the same objectives. If my interpretation is correct, then patents and the other elements of the current system of intellectual property rights are inappropriate.

The Properties of the Market vs. Intellectual Property

You might expect that the idea of public goods would seem terribly threatening to the powers that be. If economists took their theory seriously, it might be. In fact, however, the theory of public goods slips through most textbooks virtually unnoticed. In the lead article of the June 1970 issue of the *American Economic Review,* William Baumol and David Bradford began:

> The need for this paper is a paradox in itself and indeed it might be subtitled: The Purloined Proposition or The Mystery of the Mislaid Maxim. For the results which it describes have appeared many times in the literature and have been reported by most eminent economists in very prominent journals. Yet these results may well come as a surprise to many readers who will consider them to be at variance with ideas which they have long accepted.[36]

When students read about public goods in their textbooks, they usually find obscure examples, such as national defense or lighthouses, rather than items that play a more important role in the daily lives of a typical individual. Some economists find even this bland mention of public goods unsettling. Ronald Coase, a Nobel Prize–winning economist at the University of Chicago and an antagonist to government involvement in the economy, attempted to convince the profession that even lighthouses should not be treated as public goods because he found a period in which semiprivate entities provided lighthouses;[37] however, he failed to make the case that such an arrangement was particularly efficient.[38]

With the exception of a small minority, almost all economists agree with the logic of public goods. If the marginal cost is low and the fixed cost is high, markets will not work efficiently. So while conventional economics demonstrates that public goods are poor candidates for commodity status, few economists have taken notice of the fact that the rise of an information economy is thrusting the theory of public goods onto center stage.

Earlier I observed that the theory of public goods did not seem particularly threatening to most economists. This response to the theory of public goods reflects the fact that few economists recognize that public goods, rather than being the exception, are coming to play an increasingly central role in the modern economy.

The ironies associated with the economics of intellectual property go further. Following the logic of public goods, markets are inappropriate for intellectual property. Therefore, most economists now advocate policies that convert information into an object of monopoly control by treating it as intellectual property. However, practically every economics textbook goes to great lengths to show that monopoly is inefficient. In fact, monopolies are antithetical to the ideal markets that markets supposedly represent because monopolies annihilate the very competition that supposedly acts as the lifeblood of markets. This belief in competition fueled the common antagonism toward patents that I discussed in the first chapter.

Most economists make the case for awarding intellectual property rights to the "owners" of information by applying one side of the logic of public goods. They accept that in competitive markets prices fall toward marginal costs and the marginal cost of information is zero. At a zero price, firms would not have an incentive to produce information because they could not make a profit for their efforts.

Such economists conclude that the solution is to treat the information as intellectual property, thereby converting a public good into a monopoly. In making this case, they ignore the other half of the logic of public goods; namely, the central proposition of economic theory, which maintains that efficiency is maximized when goods sell for their marginal costs. Habitually caught within the narrow confines of their economic models, these economists contend that the monopoly is required to provide the incentive to create information.

Forced to choose between making information a public good or a private monopoly, these economists prefer the latter. They insist that without granting firms the monopoly privileges of intellectual property, society would be bereft of the benefits of modern technology.

Defenders of intellectual property rights try to explain their way out of this dilemma. They maintain that the flow of new efforts from firms attempting to profit from the future monopoly that intellectual property rights convey will more than offset the losses imposed by the monopoly powers of the holders of existing intellectual property rights. This assumption, as I will show, rests on questionable foundations.

In effect, then, these economists, accustomed to thinking of public goods as an exception to be relegated to a footnote in their textbooks, fail to realize that information, the ultimate public good, should not enter into the marketplace according to the logic of their discipline.

The Privatization of Public Goods

In 1954, MIT economist Paul Samuelson, later a Nobel Prize–winner, first proposed the idea of a pure public good.[39] Besides being non-rivalrous, a pure public good is not excludable, meaning that people can have access to the good regardless of whether or not they pay for the good. Samuelson's purpose was to indicate that a wide range of goods fall along a spectrum between pure public goods and those that are appropriate for private goods status.

For example, a radio signal broadcast over the airwaves is not excludable. Anyone with a receiver can hear the broadcast. In 1958, the Federal Communications Commission (FCC) was debating whether to allow the broadcast industry "to convert a public good into a private good" by permitting the scrambling of television broadcasts so that only paying customers could watch. Samuelson noted that some might think that scrambling makes public goods such as broadcasts into private goods. However, he took issue with that approach:

> Such an argument would be wrong. Being able to limit a public good's consumption does not make it a true-blue private good. For what, after all, are the true marginal costs of having one extra family tune in on the program? They are literally zero. Why then prevent any family which would receive positive pleasure from tuning in on the program from doing so?[40]

Jora Minasian responded to Samuelson in the *Journal of Law and Economics,* then edited by Ronald Coase, the economist who was so opposed to the idea of public goods. Minasian took strong issue with Samuelson's stance on pay television.[41] Minasian was concerned about who would decide what to broadcast if regulators were to regard television as a public good. How could consumers signal their preferences if not through their willingness to purchase programming?

Samuelson was troubled by this response. He raised what he considered to be a serious theoretical question in his work on public goods. Minasian had swept aside everything that Samuelson had proposed for purely ideological reasons. Indignant, Samuelson roared:

> The final question is, Why all this? Is it because despite all denials, Chicago [the university most associated with the sort of ideological reasoning that Minasian was following] is not so much a place as a state of mind? Is it because of the fear that finding an element of the public-good problem in an area is prone to deliver it over to the totalitarian state and take it away from the free market? The line between conviction and paranoia is a fine line.[42]

In effect, economists such as Minasian conceive of public goods solely from the perspective of consumption. Specifically, they identify public goods by the difficulty of excluding those who do not pay for a good or service, ignoring the fact that basic economic theory indicates that the characteristics of production, not consumption, define public goods. Adopting the perspective that Minasian proposed implies "a knife-edge pole of the private-good case, and with *all* the rest of the world in the public-good domain."[43] For this reason, Samuelson insisted on turning attention back to marginal costs.

Yes, if a public good, defined by minimal marginal costs, were to be sold in a competitive market, the price would sink toward zero. Markets cannot work under such conditions. Minasian was not concerned about that possibility because the scrambled television signal would not merely be a private good, but it would also be a private monopoly good—in effect, a form of intellectual property.

In that case, Minasian did not have to worry about that possibility of somebody else creating the same signal. His concern was that the producer be given the means to exclude nonpaying consumers, just as effectively as the producers of material goods can. Of course, scrambling the signal would not automatically convert the broadcast into intellectual property unless the legal system also were to prohibit the production of inexpensive descramblers. Minasian willfully ignored the fact, emphasized by Samuelson, that because the marginal cost of a public good is zero, charging for that good is inefficient according to the basic theory of economics.

In this spirit, those who unquestioningly defend the market regard the ingenuity that business displays in turning public goods into private goods as a positive accomplishment, one that somehow undermines the rationale for public goods.[44] They blithely ignore Paul Samuelson's point about the inefficiency of marketing public goods as private goods.

The privatization of public goods continues apace. In one of the most extreme cases, the United States awarded rights to the digital television spectrum to the broadcast industry in 1997. This transfer represented an enormous windfall, estimated by the FCC to be worth as much as $70 billion.[45]

This dismissal of the logic of public goods stands in sharp contrast to the typical approach that economists take. I feel confident that most contributors to the *Journal of Law and Economics* would be quick to invoke economic theory to rebuke the supposed ignorance of anybody who would suggest some form of government regulation that would inconvenience business. Here, when the sacred laws of economics suggest something that might not be in the best interests of business, economic theory is swept aside.

Accidental Benefits of Privatization

The case of the broadcast industry illustrates another phenomenon regarding intellectual property. In making the case that intellectual property rights create inefficiencies, I should note that even irrational or antisocial actions can ultimately provide unforeseen benefits in the long run. This assertion should not come as a surprise to a society accustomed to justifying military spending for lethal purposes, at least in part, on the basis of the spinoffs from military technology. The origins of the cable television industry nicely illustrate this point.

Cable television arose from the experience of isolated communities that had no local television stations. These communities constructed large antennae that were capable of receiving more distant transmissions. They then rebroadcast the television signals to the community at large.

Soon cable companies realized this activity was a potential source of profit, but with a major modification. Rather than rebroadcasting the signal to the entire community, they strung cables to carry the programming. These cables represented the major cost of their industry. However, the purpose of the cables was not to deliver the product, but rather to exclude everybody who did not pay the cable operator for receiving the same signal that could be inexpensively retransmitted to the entire community.

Cable companies routinely refuse to string their cables anywhere except fairly densely populated areas. Customers on the fringes were left without service until expensive satellite dishes came on the market. The cable companies make sure that their customers do not take "too much" of their product. They develop set-top boxes to prevent customers from viewing channels for which they did not pay. Because they also control most of the material available from the satellites, they installed expensive scrambling devices and descramblers so that they can degrade the signal for the satellite dish owners who do not pay for premium service.

Cable companies also incur significant costs by billing customers and connecting and disconnecting their service. Such activities do nothing to improve the service of cable companies. Today, cable companies are so profitable that when a larger company swallows a smaller one, it pays several thousand dollars per customer, even though the marginal cost of their services is trivial. In short, the cable industry has spent an enormous amount of funds to convert an inexpensive public good into an expensive commodity.

Today, cable systems are laying new and improved cables that can add to the value of their product, say by connecting customers to the Internet or carrying telephone calls. The industry never foresaw that the cables, which they originally intended merely to restrict access, could serve a positive purpose. The industry just happened upon these new uses by accident.

This unforeseen result should in no way justify the conversion of cable signals into commodities. After all, if the common application of burglar alarms turned out to serve some purpose other than protecting property from thieves, we should not credit those who threaten to rob us with providing a useful service. In addition, had the time and resources that went into creating the infrastructure of the cable system been devoted to more directly productive activities, the payoffs or spinoffs could have been many times greater than the potential benefit of the cable system.

Tactical Resistance to the Theory of Public Goods

Earlier, I made the claim that few economists seem troubled by the theoretical implications of public goods. Now, the basis of that assertion should be clearer. The idea of public goods does not exactly bother many economists, either because they willfully ignore it or because they have let themselves be misled by the typical textbook presentation of public goods as an unusual phenomenon, restricted to lighthouses and military defense.

The second group would be open to dialogue. The first group presents a more difficult challenge. Instead of worrying about how goods with low marginal costs of production cannot come to market efficiently in a competitive market, these dogmatists, following the lead of Minasian, distort the problem by framing it in terms of excludability. In other words, they sweep considerations of efficiency under the carpet, focusing their attention on the profitability of the supplier, satisfied that if business can make enough profit, nobody need worry about economic efficiency. This approach suggests a position that is inconsistent with the basic economic justification of the market, suggesting a desire to support the interests of business regardless of the consequences for the economy as a whole.

In addition, besides emphasizing the notion of excludability, conventional economists tend to frame their discussion of low marginal costs in absolute terms of whether or not goods are rivalrous. Since very few goods are absolutely non-rivalrous, this approach tends to minimize the importance of public goods, even though in a modern economy, the majority of goods probably have very low marginal costs and high fixed costs.

Samuelson laid the framework for treating such goods as quasi-public goods by insisting that goods can be more or less rivalrous, falling along a continuum. For example, if a software program costs a few cents to reproduce, it is not entirely non-rivalrous, even though it has much more in common with a public than a private good. In this respect, Samuelson showed that it should be treated as if it were a public good.

While Samuelson was correct to insist on the inefficiencies caused by treating public goods as private goods, he missed a larger dimension of the problem; namely, that the privatization of public goods can distort the nature of the goods themselves, or even the way that they are produced. For example, the scrambling of television signals creates an inefficiency that harms society, but the damage arising from this practice may seem minimal. In fact, given the quality of market-oriented television, a cynic might be tempted to remark that the scrambling represents a positive contribution by reducing the contamination of contemporary television programming. However, television technology offers a wonderful opportunity to spread information, as well as entertainment, among the population. Rather than take advantage of this potential, television has become a grotesque medium designed merely to reinforce the value of products through advertising.

In the case of science, the privatization process leads to a host of problems, including the spread of secrecy and wasteful duplication of effort. In addition, litigation associated with assigning ownership to scientific results creates a legal nightmare in which society consumes an inordinate amount of resources in fruitless litigation that is inimical to the creation of new knowledge and information. The cost of the opportunities foregone on account of these defects in the privatization of science is incalculable.

On an even deeper level, this privatization process is a major contributor to the ongoing redistribution of wealth, which consigns a large part of society to the economic underworld. While defenders of intellectual property rights admit that the monopoly of intellectual property rights causes losses today, they insist that these losses will be more than offset by the efforts of people to capture profits in the future from intellectual property rights. This concern for the future seems to have limits.

Far less attention is given to the downside of this process. As society becomes more and more stratified, relatively few people emerge from poverty with the training and experience that would allow them to make the important contributions to society that they might otherwise be able to offer. As a result, humanity loses the opportunity to benefit from much of the potential of those who are marginalized by poverty.

In opposition to the perspective that I am proposing here, defenders of intellectual property rights often present them as if they were necessary to the functioning of an efficient market. Of course, if markets really worked as well as the dogmatic advocates of laissez faire would have us think, we would have no need for intellectual property rights at all.

On Competition, Productivity, and Intellectual Property

Since competitive markets drive prices down toward marginal costs, strong competition is unsustainable in the long run because many industries have low marginal costs and high fixed costs. Although the authorities use monetary and fiscal policy to ease the competitive pressures, most competitive industries producing relatively homogeneous goods still earn very low profits, if any at all. I elaborated on this last point elsewhere.[46]

Intellectual property rights can provide a refuge from fierce competition. They also offer a competitive edge in international markets. I suspect that to a considerable degree, the upswing in profits that business has enjoyed in the past decades reflects the colossal gains that the holders of intellectual property have made. The more competitive industries have fared less well.

In this respect, modern market economies are segmenting into two tiers. Farmers, for the most part, operate on the wrong side of the intellectual property divide, selling relatively undifferentiated products while buying more and more intellectual property, such as seeds and chemicals. Between 1990-92 and 1998, the prices farmers received rose an average of 1 percent, while the prices farmers paid rose 15 percent.[47]

Much of the discussion about the so-called New Economy obscures this difference. The pundits rave about wonders of modern technology, while they tend to credit the development of this technology to the miracle of competition.

In this respect, those who spin tales of the marvels of the New Economy describe the way the system works with a worrisome lack of realism. To hear their accounts would leave the naive to conclude that the economy consisted only of the sort of high-technology firms that fight each other in highly competitive markets. In reality, the high-tech companies attempt to avoid at all costs the sort of competition that occurs in the wheat market. To a large extent, they succeed.

Virtually all high-tech firms try to do whatever is possible to avoid the sort of competition associated with markets that compete on price. According to people who follow high-tech industries, the worst fate that can befall a high-tech company is to see the product that it sells descend into the status of a commodity—meaning that its products will have to compete on price. Once that occurs, the company drops from the first to the second tier. Its expectations of future profits suffer a comparable tumble.

So, one can rhapsodize either about the marvels of competition or about the miracles of high technology, but to imagine that the two easily coexist is to strain credulity. In a market economy, high technology develops under the protective cover of intellectual property. Because of the typical quasi-public good cost structure in high technology, strong competition would be disastrous. In short, the anticompetitive properties of intellectual property rights are essential within the context of a market economy. To confuse the undeniable accomplishments of high technology with a competitive marketplace is to commit a grave error.

In industries without the protection of patents, firms try to find refuge from competition under the protection of brand names and trademarks. Those that succeed in remaining aloof from price competition earn enormous rewards. The cigarette business illustrates the importance of remaining in the top tier. In 1995, Marlboro was reportedly the most valuable brand in the world, worth an estimated $44.6 billion.[48] Even after years of bad publicity associated with successful tobacco litigation, Marlboro's brand declined only to number 11 by 2000.[49]

The power of Marlboro's marketing image is revealing. Because the tobacco industry has not been able to advertise on television for some years, programs regularly use characters from tobacco companies as corporate villains, since that industry cannot retaliate by withdrawing advertising dollars. The weakened reputation of the industry has allowed some politicians to present themselves as defenders of people against corporate power by denouncing the industry. Despite the sustained attack on the industry, the Marlboro image still retains significant value. Part of the value may be due to the addictive quality of the product, but other companies market other brands less successfully. Certainly, the enormous funds invested in the cultivation of this brand have proven successful.

The Marlboro story is instructive in showing how intellectual property can trump price competition. Profit on brand-name cigarettes runs about 55 cents per pack, while sellers of generics earn only about 5 cents.[50] Presumably, this profit differential takes into account the massive expenditures involved in saturating the world with advertising. Investors appreciate the profitability of brand names. As billionaire investor Warren Buffett, once one of the largest shareholders of RJR, the company that owns the Marlboro brand, is reported to have remarked: "I'll tell you why I like the cigarette business. . . . It costs a penny to make. Sell it for a dollar. It's addictive. And there's fantastic brand loyalty."[51]

By 2000, Coca-Cola won the honor of having the most valuable brand for its flavored water. Moreover, the value of its brand was worth more than four times 1999 sales, an achievement matched by only two other of the 75 major brands analyzed by Interbrand.[52] Here again, a product that costs very little to produce can still command a high price because of its intellectual property. Symbolic of its reliance on intellectual property, in 1986 the company divested most of its physical assets as Coca-Cola Enterprises. "Most of the value of Coca-Cola comes from the value of its secret formula and marketing know-how."[53]

Not surprisingly, company after company spends whatever seems necessary to build up its brand to protect itself against competition. Nike, for example, spent an estimated $643 million marketing its shoes in 1996.[54] In fact, marketing costs far more than production. One year, the company paid Michael Jordan $20 million for his endorsements, more than the total wage bill for all the factories in Indonesia, which were making the bulk of Nike's shoes at the time.[55] Only a handful of athletic shoe companies have the resources required to compete with Nike by creating their own brands. As a result, in this industry, competition, in terms of selling the best quality good at the lowest possible price, is virtually nonexistent.

These firms may be competing, but they are competing on brands rather than on their products. Naomi Klein collected a number of statements that indicate the corporate mindset regarding the distinction between brands and products. "Products are made in the factory," says Walter Landor, president of the Landor branding agency, "but brands are made in the mind." Peter Schweitzer, president of the advertising giant Walter Thompson, reiterates the same thought: "The difference between products and brands is fundamental. A product is something that is made in a factory; a brand is something that is bought by a customer."[56]

This effort to build up a brand often comes at the workers' expense. Nike is legendary for moving production from one low-wage economy to another one with still lower wages. John Ermatinger, president of Levi Strauss Americas division, made the connection between workers' welfare and the efforts to build up the brand in explaining the company's decision to shut down 22 plants and lay off 13,000 North American workers between November 1997 and February 1999:

> Our strategic plan in North America is to focus intensely on brand management, marketing and product design as a means to meet the casual clothing wants and needs of consumers. Shifting a significant portion of our manufacturing from the U.S. and Canadian markets to contractors throughout the world will give the company greater flexibility to allocate resources and capital to its brands. These steps are crucial if we are to remain competitive.[57]

Over and above brand names, intellectual property rights provide a powerful form of market power that increases the value of commodities. Falling computer prices tend to distract the public from this phenomenon. In terms of computer hardware, where intellectual property rights play less of a role, prices have plummeted, allowing the users of computers to increase their productivity. Even here, intellectual property rights add to the cost of a computer, but rapid technological progress has masked these price increases.

Computer hardware is the exception that proves the rule. In most industries, technological progress has not been sufficient to offset the costs that intellectual property rights impose on the rest of society. The hefty price increases imposed by the pharmaceutical industry are closer to the norm. To the extent that these intellectual property rights allow a pharmaceutical company to increase the price it charges for a pill, the measured productivity of that business will increase. In effect, then, intellectual property creates the illusion of productivity gains.

At the same time, businesses that have to pay the costs of intellectual property rights to the companies that supply them suffer lower profits and the appearance of less productivity. These symptoms tend to encourage the government to pamper the industries with strong intellectual property rights while treating other parts of the economy less generously.

CHAPTER SIX

THE COSTS OF INTELLECTUAL PROPERTY

How Much Does the System of Intellectual Property Rights Cost?

I have repeatedly mentioned the incalculable resources that the present system of intellectual property rights consumes. Let me get more specific. Frederic Scherer, of Harvard's Kennedy School and formerly director of the Federal Trade Commission's Bureau of Economics, is arguably the foremost economist today on the subject of industrial organization. He calculated an estimate for the cost of the patent system back in 1978, when the current system was in its infancy. Recall that the new court to promote intellectual property began in 1980.

Scherer began with the total budget of the U.S. Patent and Trademark Office for fiscal year 1978, which was then $90 million, including salaries for 2,945 authorized employees plus other sundry expenses. The American Patent Bar Association has nearly 3,000 members, most of whom worked outside the Patent Office. Assuming salary, clerical support, transportation, and office maintenance costs of roughly $80,000 per practicing patent attorney, he estimated total annual patent system administration costs of roughly $330 million at 1978 prices.[1]

Of course, Sherer's estimate is more than two decades old. Since then, the size of the Patent and Trademark Office has grown enormously. According to its 1998 annual report, the latest posted on the Web in June 2000, the agency planned to hire an additional thousand people in 1999 alone.[2]

The current system of granting monopoly rights on the basis of a patent creates additional costs that are impossible to measure. Companies devote considerable effort to create patents, even when they might be of no direct use for them. Instead, each firm wants to be able to sue for royalties and/or to accumulate a strategic portfolio of patents with which to bludgeon other companies that might challenge them with an intellectual property suit in the future.

Once a company wins a patent, other companies begin to organize their business around the activities of the company that initially won the patent. Soon, a large infrastructure, including suppliers and distributors, can develop around that patent. After a short time, this structure of businesses organized around a patent becomes more or less locked into place.

This sort of economic reorganization is not unique to the patent system, but the effects of the patent system in this regard can create substantial costs. Presently, the patent system is expanding headlong into areas without much thought about the implications of these new policies. Certainly, a good number of mistakes will be made in awarding patents. Reversing such a mistake can impose profound costs upon the whole structure of businesses built up around the improper granting of a patent.

Since intellectual property protection depends on the laws that Congress passes, lobbying for advantages in the system of intellectual property represents an enormous, but uncountable, expense. In addition, think tanks and influence peddlers are begging to devote considerable time and energy to sway the thinking of judges. One common practice is to invite them to a "seminar" at a swank resort.

Scherer also left out any consideration of the time and expense business consumes in circumventing patents, through techniques such as reverse engineering. Such efforts to work around existing patents represent an enormous, but unmeasurable, cost. Recall the discussion of the efforts to reverse-engineer computer chips and duplicate the qualities of pharmaceuticals.

Ignore these added costs, which may well exceed Scherer's estimate. Then compare that not insignificant sum with Fred Warshofsky's estimate of the cost of merely searching for possible conflicts with existing patents for a single piece of software:

> A [software] system with 100,000 components, for example, can use hundreds of previously patented techniques. Because each patent search costs about a thousand dollars, searching for all the possible patent potholes in the program could easily run well over $1 million, and that far exceeds the cost of writing the program.[3]

Writing in the early days of biotechnology, a *Business Week* article cited an industry attorney who observed, "there are more lawsuits than products." At the time, Genentech sold only two products; it was already engaged in seven suits.[4]

The American Intellectual Property Law group estimated that the median cost of a single patent case in 1994 was $280,000 for each side through discovery and $518,000 through trial.[5] By 1998, the average cost of a trial had soared to about $1.5 million per side.[6]

During the early 1990s, Intel's annual litigation budget alone was believed to be at least $100 million—or about one-third of Scherer's estimated cost of the entire system back in 1978.[7] No doubt, Intel's litigation budget has grown substantially since then. In addition, any estimate of the costs of the present system of intellectual property rights would have to include a portion of the expense of maintaining the judicial system—especially since the establishment of a federal court for the sole purpose of hearing patent cases.

While the expense of protecting intellectual property rights may be high, the benefits for a holder of intellectual property rights may be immense. A firm that can bring a patent suit to exclude competition, even for a relatively short time, may enjoy phenomenal profits. For example, a successful drug generates about $1 million in sales per day.[8] Obviously, excluding a competitor through costly legal maneuvers, even when the patent claim lacks merit, might be an attractive strategy when the profits from a delay are so high.

As the stakes in intellectual property disputes increase, corporations turn to giant legal firms noted for their political connections as well as their legal expertise. For example, according to a June 30, 1999, press release of Akin, Gump, Strauss, Hauer & Feld, that firm's merger with Panitch Schwarze Jacobs & Nadel brought together two firms that obtained a combined total of 659 patents for their clients in 1998. This firm is not just big; it is politically powerful, employing many people who previously occupied positions in the highest levels of government. The Center for Responsive Politics estimates that in 1998 alone this firm hauled in $11.8 million in lobbying fees.[9] This sort of influence will undoubtedly be useful in matters of high-value intellectual property rights, as well as in shaping legislation that will favor specific holders of intellectual property. In this manner, the patent system contributes to the general contamination of the political process.

In this litigious world in which influence looms so large, the independent inventors, and even small corporations, find themselves at a serious disadvantage. To the extent that this disadvantage discourages independent inventors and small corporations, the present system of intellectual property imposes still another cost to society as a whole.[10]

As I have noted earlier, intellectual property rights are to science what tollbooths are to highway traffic. Both create bottlenecks and impede forward progress, but in the case of intellectual property rights, innumerable disputes arise about who gets to collect the tolls and how much the tolls should be. To the extent that the present system of intellectual property constricts the flow of new technologies, it imposes another incalculable cost on society.

Consider for the moment not just the patent system, but intellectual property in general. The costs go far beyond the costs of the system that I have already mentioned. For example, recall the earlier discussion about the relationship between the role of the U.S. military and the defense of intellectual property.

How much of the military budget reflects the need to defend intellectual property? Such questions are unanswerable, but important nonetheless. Similarly, a certain portion of the overall criminal justice system goes to the protection of intellectual property rights. Again, nobody can accurately calculate how much of this enormous expense reflects the protection of intellectual property, but certainly it represents a significant sum.

Maybe Even Greater Costs of Intellectual Property Rights

Taking the argument one step further, since intellectual property frequently reflects ideas and information rather than material objects, the protection of intellectual property is much more difficult than the protection of more conventional forms of property. For example, if I take a can of soup from the store without paying for it, security guards may catch me in the act.

The physical aspects of the unauthorized acquisition of information are more ambiguous than the shoplifting of a can of soup. As a result, zealous protection of intellectual property rights to information requires a far more intrusive invasion of the public's privacy than does safeguarding conventional property rights.

To make matters worse, the present system even protects information about information. For example, the oppressive Digital Millennium Copyright Act forbids the circulation of information about methods that might be used to defeat measures to protect intellectual property.

Let me give you an example of how this works. In September 2000, a consortium affiliated with the music industry known as the Secure Digital Music Initiative, or SDMI, invited volunteers to test the security of its antipiracy technologies, known as watermarks, which are embedded codes in digital-music files that can be used to block copies.

A team led by Princeton University computer science professor Edward Felten accepted the challenge and cracked the code. Besides defeating four of the watermarking methods in the challenge, Felten's group was able to make educated guesses about how the watermarking was done in the first place.

The team had planned to present its results in April 2001 at the International Information Hiding Workshop in Pittsburgh. Before the conference, Matthew Oppenheim, head of the SDMI Foundation, sent Felten a letter warning him to withdraw his conference paper and to ensure that it would be removed from the conference proceedings and "destroyed." Failing to do so could potentially expose the scientists to legal action under the Digital Millennium Copyright Act.[11] In short, copyright protection trumps freedom of speech.

In another case, a 16-year-old Norwegian boy, Jon Johansen, wanted to use DVDs on his Unix system, but the Content Scrambling System (CSS), created for the Motion Picture Association of America to prevent piracy, did not allow legal use of DVDs on a Unix system. The Norwegian government charged Johansen with an economic crime because he wrote some simple code, known as DeCSS, to be able to run DVDs on his Unix system. Later, the Motion Picture Association of America successfully sued the magazine *2600* for informing people where they could find the DeCSS code, although the decision is now under appeal.

Judging by their behavior, the holders of intellectual property rights expect to be able to establish their complete control over how people use what they claim as their property. While this claim may seem excessive, let me give a few examples.

Just as Monsanto no longer sells seeds, Microsoft now wants to stop selling its software. Instead, it wants to lease it for a limited time. Presumably, it will install code to disable the program after the lease expires. To override the code would constitute a crime under the Digital Millennium Copyright Act.

Adobe recently released an e-book of *Alice in Wonderland.* The book in itself contains no intellectual property on the part of Adobe. In fact, the company took the text from the marvelous Project Gutenberg, which freely distributes electronic copies of books from the public domain at no cost. The very first section of the Adobe book contains a list of "permissions" regarding its use. In the list of "permissions," the company specified: "this book cannot be lent or given to someone" and "this book cannot be read aloud." What penalty would be imposed on a parent who defiantly read this book to a child? The company modified its permissions after it came under criticism, but the fact remains that these corporations seem to see no limit to the amount of control that they can impose on people's activities to protect their intellectual property.[12]

Adobe is not alone in its efforts. Pat Schroeder, once known as a liberal member of the U.S. House of Representatives, is now CEO of the Association of American Publishers. She has been quoted as "saying that publishers have to 'learn to push back' against libraries, which she portrays as an organized band of pirates!" Peter Chernin, president and chief operating officer of Rupert

Murdoch's News Corporation, which counts HarperCollins Publishers among its vast media holdings, is calling for legislation that "guarantees publishers' control of not only the integrity of an original work, but of the extent and duration of users' access to that work, the availability of data about the work and restrictions on forwarding the work to others."[13]

Let me return now to the question of the DeCSS suit. Keep in mind that the magazine had not even distributed the code; it had merely let people know where they could find the code. The code in dispute, DeCSS, is the same program that I earlier mentioned in passing while discussing the ridiculous lengths to which the government is willing to go to justify intellectual property rights. You may recall how Assistant U.S. Attorney Daniel Alter compared DeCSS to other tools that terrorists might use.

In this kind of climate, human rights of all kinds must give way to intellectual property rights. This tragic cost of intellectual property rights might exceed all other costs put together.

Intellectual Property and the Redistribution of Wealth

The ability to earn monopoly profits from intellectual property rights redistributes wealth from those who purchase the good to those who sell it. Certainly, intellectual property is a major factor in creating the grotesque rise in inequality occurring in the United States today.

Only on rare occasions have economists taken the time to study the effect of monopoly on the distribution of income. When they do analyze the relationship, they find a profound effect on the degree on inequality. For example, William S. Comanor and John Smiley, writing in 1975—a time when the distribution of wealth and income was far less concentrated than today—estimated that the share of wealth of the 2.4 percent of the households that then held more than 40 percent of total wealth would fall to 32 percent in the absence of monopoly. Even more dramatically, the share of 0.27 percent of the households that then held more than 18.5 percent of total wealth would fall to 13 percent in the absence of monopoly.[14]

Writing little more than a decade later, Irene Powell, an economist who was at Mount Holyoke College, analyzed the effect of monopoly in the manufacturing sector. She found that in addition to transferring income from the public at large to the owners of the corporate sector, monopoly extracted a larger share of income from the lower sectors of the income distribution.[15]

The phenomena that captured the attention of these economists were quite modest compared to what is underway at this time. The situation today is very similar to what occurred during the Age of the Robber Barons in the late nineteenth century, with one exception. Many—but not all—of the robber barons earned their wealth in the process of constructing the infrastructure that made possible the economic development of the twentieth century.

In contrast, the majority of the more recent great fortunes—especially those associated with intellectual property—result from the transfer, rather than the production of wealth. Even people who occupy the upper strata of society are coming to realize the nature of this process. For example, John Doerr, one of the most active venture capitalists in Silicon Valley, reflected this distinction between the earlier transfer of wealth and its contemporary counterpart when he referred to digital technology as "the largest legal creation of wealth in the history of the planet."[16] Even more directly, Warren Buffett, the wealthiest investor in the United States, told the 15,000 shareholders at the annual meeting of Berkshire Hathaway, "When we look back, we will see this as a period of enormous amounts of wealth transfer [rather than wealth creation]."[17]

Of course, pushing the distinction between the robber barons and the modern holders of intellectual property too far would be a mistake. Certainly, some of the robber barons had nothing at all to do with the development of industry. Instead, they simply defrauded investors, while redirecting corporate resources into their own private investments. In addition, some of the contemporary holders of intellectual property are no doubt also making positive contributions to the economy. Even so, the redistributive nature of the bulk of the great fortunes of today still holds.

Many of the costs of intellectual property fall within a category that economists refer to as rent-seeking, meaning the wasteful dissipation of resources in an effort to be able to collect income that is unearned in any meaningful sense of the word. This rent-seeking takes many forms. Corporations devote substantial resources to litigation, lobbying, public relations, reverse engineering, advertising, and other activities aimed at acquiring or strengthening intellectual property rights. None of this activity is productive by any stretch of the imagination.

As the rewards for intellectual property escalate, you can be sure that the rent-seeking efforts of corporations will skyrocket even faster. Unless something is done to correct this situation, I suspect that Charles Dickens's classic *Bleak House* might come back into fashion. In that book, Dickens, the zealous defender of intellectual property, provided a wonderful metaphor of the horrors of litigation about intellectual property. He portrayed opposing lawyers fighting so long over a disputed inheritance in Jarndyce v. Jarndyce that legal fees consumed the whole amount.[18]

The massive concentrations of wealth associated with intellectual property encourage ostentatious displays of affluence. Competition in constructing trophy mansions or purchasing other symbols of success becomes commonplace. In this environment middle class people get caught up in a whirlwind of consumption.[19] They feel pressure to spend a good deal of money to signal that they are successful to potential clients or to live in a neighborhood that provides good schools for their children, which puts strong lifestyle demands upon them.

Intellectual Property and the Disruption of Society

According to the infamous trickle-down theory, the accumulation of wealth by the holders of intellectual property should eventually bring prosperity to less affluent members of society. If that theory were correct, then the poor in California, which probably has the highest concentration of intellectual property in the world, should be doing quite well. In fact, their condition has deteriorated substantially. One study found that

> between 1969 and 1997, real wages for male workers [in California] grew only at the very top. . . . At the median and below, male wages declined dramatically and steadily. . . . In other words, while the rich got a little richer, the poor got a whole lot poorer. To be more precise, male workers who were at the bottom of the distribution in 1997 had even lower wages than the male workers at the bottom in 1969.[20]

Another study, by researchers at the Federal Reserve Bank of San Francisco, concluded that

> after six years of solid economic growth, a larger number of Californians are living in poverty, a smaller number are in the middle class, and a majority have family incomes below those observed in 1989, the last business cycle peak. Moreover, a majority of families in California have less income than comparable families living elsewhere in the U.S.[21]

The trickle down, however, is not absolutely ineffective. For example, salaries for first-year lawyers in Silicon Valley are skyrocketing. Firms are offering recent graduates of law school first-year salaries of $125,000 with a guaranteed bonus of $20,000 and the possibility of another $5,000.[22]

For less advantaged elements of society, the trickle down does nothing to improve their fortunes. Instead, in the case of the strengthening of intellectual property rights, the results are increased impoverishment of the poor.

The association of intellectual property with poverty might seem far-fetched at first glance. References to the New Economy evoke images of highly trained scientists or programmers who develop new genetic strains or assemble sophisticated programs. True, the young girls in Chinese or Haitian sweatshops who manufacture Nike sneakers or Disney shirts live in abject poverty, but what does their situation have to do with workers in Silicon Valley? Isn't Silicon Valley worlds apart from the sweatshops in the poorer countries?

In fact, the two worlds of intellectual property are not as far apart as they might seem. The early history of IBM computers illustrates the similarity. In the mid-1960s, the company could not keep pace with demand. It developed a high-tech, automated factory in Kingston, Colorado, to build memory systems for its computers, but it still could not keep up with demand. One executive had some experience working with IBM in Japan. He sent some materials to Japan to have poor Japanese women assemble the memory systems by hand. IBM was surprised to find that the quality was as good as the output from its modern plant and the cost was considerably lower.[23] Soon, reliance upon Asian production became widespread in the electronics industry.

Even today, where the electronics industry produces goods domestically, it still relies on low-wage labor. As a result, many workers in the high-tech industries do not share in the prosperity of their employers.

Homelessness in the Home of the Modern Electronics Industry

Consider housing for an example. The press occasionally reports on leading moguls competing with one another to build the most ostentatious mansions. In the regions where intellectual property wealth is most comfortable, lesser lights must content themselves with finding comfortable housing within an easy commute of their work, since traffic jams are one of the few democratic institutions remaining in the United States—rich and poor alike must wait in snarled traffic. In fact, *The Economist* estimates that drivers in Santa Clara County, in the heart of Silicon Valley, lose approximately 30,000 hours every day. Taken over an entire working year of about 250 days, the time lost in traffic jams would be the equivalent of 3,750 full-time employees.[24]

The concentration of wealth, however, allows the wealthiest a certain exemption from traffic by bidding housing away from less fortunate citizens. This process widens the gap between rich and poor by reducing the standard of living of the poor. This rush to acquire property, either for ostentation or convenience, might seem harmless enough, but to the previous residents of the area it can mean wrenching disruption of their lives. Since the wealthy prefer to build expansive houses on larger properties, which would otherwise be suitable for multiple dwellings, the region added only 54,600 housing units between 1992 and 1999, while it created 275,000 jobs.[25]

As the wealthy bid property away from the less fortunate, the term "affordable housing" becomes a tragic oxymoron for all but the most affluent. The result is a form of economic apartheid. Housing prices became so high in Silicon Valley that a family of four scraping by on $53,100 a year or an individual earning less than $37,200 was officially poor according to federal housing officials. Indeed, the median price for a house in Silicon Valley, $410,000, became more than twice that for the rest of the country.[26]

While almost unimaginable wealth flowed to some of the residents of this seeming dot.com paradise, average incomes for the lowest fifth of valley households actually fell during most of the 1990s.[27] In the words of Amy Dean, the energetic director of the AFL-CIO's Silicon Valley office: "Unfortunately, the New Economy is looking a lot like an hourglass with a lot of high-paid, high-tech jobs at the high end and an enormous proliferation of low-wage service jobs at the bottom."[28]

Less than 30 percent of the households in Santa Clara County can afford to buy a house. In San Francisco, rents became even more unaffordable. There, the median price of a home rose to close to $500,000, a price that only 11 percent of the city's 750,000 residents can afford.[29]

Renting is increasingly out of range for the average worker as well. Two out of five valley residents can afford to rent the average two-bedroom apartment, which runs about $1,700. A person earning the minimum wage in San Francisco, devoting the entire salary to rent, would have to work the equivalent of 174 hours to afford the median rent in that city.[30] No wonder, then, that even some high-tech workers end up on the church soup lines. In fact, in Santa Clara County, 34 percent of the estimated 20,000 homeless people in 1999 had full-time jobs, up from 25 percent in 1995.[31]

These figures fail to count the growing number of families doubled up in single apartments, or paying $400 a month to live in a garage or to sleep on a stranger's living-room floor.[32] In addition, many who do find housing do so far from their place of employment. As a result, they must endure very long commutes in dense traffic, which consumes a good part of the day away from the job.[33]

This problem is not unique to the San Francisco Bay area. Indeed, the explosion in the number of full-time workers in poverty is a national disgrace. Since 1986, the ranks of full-time workers in poverty have grown by an astounding 40 percent.[34] This phenomenon seems to be associated with intellectual property. For example, the U.S. Department of Housing and Urban Development found that rapid rent increases were common among those regions enjoying the fastest rate of employment growth in high technology.[35] In Fairfax County, Virginia, a wealthy suburb of Washington, D.C., for instance, 64 percent of the homeless are working.[36]

Obviously, some of this disparity between the rich and the poor reflects the impersonal workings of the market. In addition, the monopoly position of holders of intellectual property rights has facilitated the transfer of wealth and income. Not content with these advantages, many of the most prominent owners of intellectual property have intentionally acted to keep the poor down. To begin with, they take extraordinary measures to pay their workers as little as possible. Often they turn to poor, uneducated workers who labor anonymously amidst great fortunes.

In order to avoid direct responsibility for their actions, these owners of intellectual property contract out their work to small-time contractors. In that fashion, they can deflect any possible public outrage about abuses of power onto the contractors, while claiming ignorance about the situation.

Another common tactic is to hire workers through temporary agencies so that the companies do not have to pay benefits. These corporations even hire highly skilled workers as temporary workers. Employees often accept this status hoping that they can somehow become permanent workers enjoying the benefits of stock options and the like.

As the accumulation of great fortunes associated with intellectual property displaces the less fortunate workers, the disruptions in their children's lives is sure to create further deleterious consequences that will reverberate in future generations. Sadly, society's real intellectual heritage—the minds of the young—will be left to wither in homeless shelters, while the supposed creators of society's intellectual property enjoy their newfound riches.

Reckonings

Much of the wealth captured through intellectual property rights moved through the stock market. Companies recruited specialists by offering them generous stock options in lieu of high wages. Holders of intellectual property rights floated new issues of stock. Gullible investors, seeing the almost miraculous gains of earlier investors, threw money at projects with only the remotest chance of success.

Eventually, the bubble burst. By late March 2001, an estimated $4.6 trillion worth of stock market value had evaporated.[37] Billionaire intellectual property barons, such as Jay Walker, watched while their paper empires tottered. The *Wall Street Journal* reported that about 80 percent of the Bay

Area Web firms were expected to fail, threatening to wipe out some 30,000 jobs. Great swaths of office space in the Bay Area suddenly became vacant.[38] Home prices in the region also fell rapidly.[39]

John Doerr, the venture capitalist, publicly apologized for contributing to the speculative frenzy with his previous claim about the enormous wealth-creating capacity of the Internet while asking a group of congressmen to grant tax relief to the speculators by allowing companies a one-year write-off of technology investments.[40]

While stock prices returned to earlier levels, this reversal could not help restore the lives of those whose lives were disrupted to make way for the spectacular influx of wealth and prosperity that accompanied the rise of the dot.com economy.

The resources, both material and human, that people threw into the speculative frenzy associated with the rise of the dot.com economy were of incalculable value. Yes, a few of these companies will survive. Once a tiny number of these survivors join the ranks of the great multinational corporations, some people might look back at the spectacle with approval. After all, markets have selected the winners in this way for a long time. For example, decades ago literally hundreds of automobile companies competed in the United States before the market whittled the field down to a handful.

A more objective analysis paints a different picture. Keep in mind that virtually anything that a Web company does can be duplicated rather cheaply, so this wild rush for wealth revolved over who could capture intellectual property rights. Had society developed a more orderly method for developing new technologies, much more could have been accomplished.

I am thinking about how scientists used to practice before they got caught up in intellectual property. I am thinking of an ethic comparable to Jonas Salk's, with his disdain of patents.

How Corporate Power Affects Intellectual Property

The new regime of intellectual property has set a dangerous process in motion. Over and above exacerbating the maldistribution of income and wealth and reinforcing the powers of corporations, it changes the entire political, social, and legal environment in a way that makes this bad situation worse.

Think back to the earlier age in which Abraham Lincoln rhapsodized about the importance of the individual inventor. Lincoln's sincerity in this regard might be open to question. After all, as an attorney, Lincoln represented the railroads, which were notorious violators of the intellectual property of independent inventors. Nonetheless, the patent law at the time embodied a defense of the rights of the independent inventor vis-à-vis the corporations.

Once the courts bowed to the growing power and influence of the railroads and the other major corporations, the courts granted corporations all the constitutional rights of an individual person. Suddenly, as described earlier, corporations began to amass an enormous array of intellectual property rights, more often than not for anticompetitive purposes.

Within this environment, the profits associated with intellectual property soared, as did the income and wealth of those who controlled the corporations. The standard used to gauge the appropriate reward for the owner of intellectual property no longer was that of the modest income of a well-paid professional, but rather an income hundreds of times more lucrative than what an ordinary worker can earn.

Given the enormous stakes involved, the major players in this game began to use their wealth and influence to subvert the political process in an effort to make the rules even more favorable to their own interests. In addition, they also gained control of the major organs of science, including much of higher education.

As a result, the hypothetical individual, who purportedly is the beneficiary of the system of intellectual property, becomes irrelevant. Supposedly, the individual has an interest in the system as a consumer, who is expected to benefit from the flow of new science and technology. Unfortunately, as I have shown, in the long run, the current system of intellectual property is probably a hindrance rather than an incentive to the development of improved technologies.

Intellectual property rights introduce still another perversion into modern society. Supposedly, what justifies the enormous gap between the income of the Silicon Valley executive and the janitor who cleans the offices for minimum wage is the difference in their technical skills. To the extent that the Silicon Valley executive's pursuit of intellectual property rights raises the price, the janitor's children will have less exposure to computers or software. By stunting this aspect of their education, the gap between the classes will become more extreme.

The government, the press, and even industry leaders bemoan the corrosive influence of the digital divide between the rich and poor, but to the extent that they applaud the strengthening of intellectual property rights, they contribute to the ever-widening digital divide. In addition, to the extent that the government musters resources for the protection of intellectual property rights, budgets for pressing social needs will be neglected, further diminishing the opportunities of the less privileged sectors of society.

The Law of Unintended Consequences

The idea that public intervention might improve the development of technology runs counter to the widespread distrust of government intervention. Today, trust in the efficacy of Adam Smith's invisible hand has become an article of faith for many economists.

Smith proposed that individual people seeking their own advantage create outcomes quite unlike what they had intended. Unfortunately, economists, following in this tradition, unquestionably accept that in a competitive market the result will be efficient without giving much consideration to the special conditions that must hold for that theory to have any validity.

For example, Smith's theory assumes away behaviors that can affect other people, except through the operation of the price system. Consequently, in a Smithian world, firms do not emit pollution that affects people in any way. Nor do scientists perform research that affects the value of what other scientists are doing. In a Smithian world, either no long-lived capital goods exist or, if they do, then investors have sufficient information about the future to allow them to invest efficiently. Finally, public goods do not exist in the context of Adam Smith's economic theory.

In recent years, as economics became more mathematically sophisticated, a few economists could no longer reconcile the sterility of economic models that required such extreme assumptions. One critical theory that gained wide circulation concerned the history of the layout of the simple typewriter keyboard as told by Paul David, the exceptionally insightful economic historian at Stanford University to whom I referred in discussing the history of scientific societies.

The first part of David's analysis is relatively uncontroversial. Early typewriters tended to get stuck if the typists struck a sequence of keys without a sufficient delay between each individual keystroke. As a result, manufacturers produced a typewriter keyboard intentionally designed to slow down the typist. The controversy begins at the point where David insists that, as typewriters became more efficient, the need to slow down the speed of typing disappeared, but the layout of the keyboard remained unchanged. The sheer inertia of this technology was sufficient to prevent the industry from manufacturing keyboards that could have offered substantially more productivity.[41]

Soon thereafter, Brian Arthur, a Stanford colleague of David's, published an article that won him broad acclaim in the media. Arthur extended the keyboard story to a number of different technologies. According to Arthur, industry commonly gets locked into technologies that might seem to be the most efficient at the time of their adoption, but which, in the long run, represent an inferior technological choice.[42]

Arthur's work echoed the iconoclastic economist Thorstein Veblen, who, a half century before, had attributed Britain's loss of economic leadership to its earlier industrial success. Veblen argued that when Britain first industrialized, it created a complex network of industrial structures. As a result, replacing any single part of this structure became very expensive, since it required further changes in other parts of the network. This inertia left Britain with factories and transportation systems that could not accommodate more modern technologies.[43]

Newly industrialized nations, such as the United States, faced no such difficulty. They were able to adopt the most modern technologies without having to incur such supplemental expenses. Thus, the British were "paying the penalty for having been thrown into the lead and so having shown the way."[44]

While Veblen's work elicited only a few interested comments, the more modern version of this theory provoked a firestorm of controversy among economists.[45] In effect, rather than accepting the possibility that the market might not always lead business to move the economy in the most efficient direction, conservative ideologues reverse the argument, invoking the principle of unintended consequences. They insist that any public actions taken in an effort to control the economy are more likely to do harm than good.

The conservatives caution the public to abandon all efforts to control its destiny and allow the wisdom of the marketplace alone to decide what is best for society. In effect, then, markets lead to good unintended consequences, while government actions lead to bad unintended consequences. An editorial page article in the *Wall Street Journal* singled out Brian's Arthur's theory for laying the judicial groundwork for the government's antitrust suit against Microsoft. The author also took strong exception to Arthur's insinuation that Microsoft's software might not be of the highest quality.[46]

Two points are worth noting here. First, intellectual property rights are antithetical to the ideal of markets. More important, neither intellectual property rights nor the market will nurture a society in which science and technology can flourish. The net effects of the many costs of intellectual property rights detailed in earlier chapters are certainly destructive, but pale in comparison to the human costs addressed in this chapter. The combination of intellectual property rights with free market ideology run amok is lethal for the less privileged sectors of society, especially the children. The loss of the potential genius and creativity of the many children who will mature in an environment of hopelessness will exact an enormous toll in the future and represents a chilling indictment of the current state of affairs.

CONCLUDING THOUGHTS

Overview

I have intended this book to raise questions rather than answer them. Toward this end, I have addressed a multitude of problems associated with intellectual property. In the process, I hope to have shown that the current regime of excessively strong intellectual property rights represents a confiscation of creativity, which undermines science, corrodes the university environment, and creates an explosion of litigation. Given the present climate, in which holders of intellectual property have undue influence over the political process, the situation can only become worse without strong countermeasures.

In addition, intellectual property rights warp the economy, especially by accentuating the maldistribution of income. At the same time, the monopoly powers granted by intellectual property undermine competition, which is supposed to be the lifeblood of a dynamic economy. Moreover, attempts to circumvent these monopolistic privileges through activities such as reverse engineering and modifying research and development in an attempt to avoid existing intellectual property rights dissipate enormous time and effort. All of these effects will cause the rate of economic growth to deteriorate.

You have seen that when the military views a particular technology as a vital resource, it does not trust its development to the market, as the history of radio and aircraft clearly shows. Instead, it takes a cavalier approach to intellectual property, knowing that the otherwise inevitable litigation would prove to be destructive.

Unlike this book, most commentaries portray intellectual property rights as reasonable. Indeed, most people understandably appreciate the incredible array of new technologies made available to the public almost daily. Given the continuing flow of new technologies made available to those who can afford them, many people naturally flinch from any suggestion to interfere with the current system.

I have attempted to show that the real threat to rapid technological progress comes from the intense application of intellectual property rights rather than from any challenge to intellectual property. In fact, the most important of modern technologies depend upon creativity from a time that predates the recent revolution in intellectual property rights. Public investment funded most of this fundamental research. In addition to public science, the public at large has built the educational system that created the infrastructure that made modern science possible. Nonetheless, private interests confiscate most of the fruits of this public investment, claiming their reward on the grounds of their ownership of intellectual property.

Because intellectual property rights must be protected at all costs, the government mobilizes the resources of the military and judicial arms to defend the holders of intellectual property rights. Even personal liberty must give way. I am sure that the catalogue of ills covered in this book is far from exhaustive. Every day, the press reports on some new and unexpected problem created by intellectual property.

Possibilities for the Future

At the end of a book like this, the author is usually obligated to offer some fairly simplistic reforms. Of course, the scope of both patents and copyrights should be far more restrictive than it is today. Michael Kremer's auction system, which I discussed earlier, might provide some improvement. Such modest reforms might remove some of the most flagrant abuses associated with intellectual property, but they would not go very far.

I would also recommend the promotion of public science by giving far more funding for basic research. The majority of public science performed in the United States probably has been directed toward serving military purposes or, to a lesser extent, the needs of specific industries.

These modest recommendations are obviously inadequate. The problems associated with intellectual property rights are so pervasive that much more sweeping actions are necessary. I cannot pretend to have all the answers to the challenges that intellectual property poses. Nobody does.

I can say with certainty that if ideas and information represent the core of a modern economy, then markets are an inappropriate mechanism for organizing such activity. As I have already shown, ideas and information are, by their very nature, public goods. Rationing them by the market makes no sense whatsoever.

I can understand the rationale for granting intellectual property rights for very specific inventions—say a new kind of wrench—but society will certainly do better to rely on public institutions rather than monetary incentives for basic scientific discoveries. Other inventions—what economists call general purpose technologies—have such far-reaching ramifications for society that they should not be left to the market.[1]

I would not dare to suggest a detailed blueprint for the organization of science. Certainly, nobody has ever discovered any single correct way of organizing the scientific process. I do feel confident that institutions in which people are free to explore matters that they find inherently interesting will foster rapid technological progress. Alternatively, necessity can provide an effective stimulus. Recall the enormous scientific and technological achievements associated with military research during World War II. I suspect that situations of necessity are more conducive to improvements in applied technology, while conditions of freedom would be more likely to support pure science.

If the goal is to unleash the creative powers of science and technology to make a better society, then the transformation of the fundamental economic institutions should rank among the highest priorities in society today. To carry out this transformation in an appropriate manner will require the active participation of as many people as possible from every strata of society.

Hopefully, by calling attention to the challenge that intellectual property poses, people will be encouraged to join in a movement to rein in intellectual property. Of course, serious though the problems associated with intellectual property are, I do not pretend that intellectual property is the only problem challenging society today, nor even the most pressing one. Nonetheless, I feel confident that once people organize to meet the challenges that intellectual property pose, they will be better able to take on many of these other problems. Information, together with a determined political will, is a basic requirement for success.

NOTES

Introduction

1. David T. Bazelon, *The Paper Economy* (NY: Random House, 1963), p. 64.
2. Peter Newcomb, "The Richest People in America," *Forbes* (October 11, 1999): 169.
3. Michael J. Mandel, "How the Super-Rich Lucked Out Twice: New Data Show the Top Earners Are Already Enjoying Lower Rates," *Business Week* (May 14, 2001): 52.
4. Erika Brown, Doug Donovan, Joanne Gordon, and Peter Newcomb, "Global Billionaires," *Forbes* (July 5, 1999).
5. Bittlingmayer and Hazlett 2000 coined the expression DOS Capital. See George Bittlingmayer and Thomas W. Hazlett, "DOS Kapital: Has Antitrust Action Against Microsoft Created Value in the Computer Industry?," *Journal of Financial Economics,* Vol. 55, No. 3 (March 2000): 329-59.
6. Newcomb, "The Richest People in America," p. 169.
7. United Nations Development Programme, *Globalization with a Human Face: United Nations Human Development Report* (NY: Oxford University Press, 1999), p. 68.
8. Karl Marx, Capital, Vol. 1 (NY: Vintage, 1977), p. 496.
9. Michael A. Perelman, *Class Warfare in the Information Age* (NY: St. Martin's Press, 1998).
10. *Davoll v. Brown,* 7 F. Case. 197 (Circuit Court, D. Massachusetts 1845).
11. *Mitchell v. Tilghman,* 86 U.S. 287.
12. Supreme Court decision 1949: *C.I.R. v. Wodehouse,* 337 U.S. 369.

Chapter One

1. Robert P. Merges, "The Economic Impact of Intellectual Property Rights: An Overview and Guide," *Journal of Cultural Economics,* Vol. 19, No. 2 (1995): 106.
2. Edith T. Penrose, *The Economics of the International Patent System* (Baltimore: Johns Hopkins University Press, 1951), p. 2.
3. Daniel Defoe, *A Plan of English Commerce* (London: C. Rivington; Kress Goldsmith Collection, Reel 407, No. 6594, 1728), pp. 298-300.
4. Merges, "The Economic Impact of Intellectual Property Rights," p. 106.
5. Fritz Machlup and Edith Penrose, "The Patent Controversy in the Nineteenth Century," *Journal of Economic History,* Vol. 10, No. 1 (May 1950): 1.
6. Ibid., p. 5.
7. Ibid.
8. Charles Dickens, "Letter to John Forster (February 24, 1841)," *Dickens on America and the Americans,* ed. Michael Slater (Austin: University of Texas Press, 1978); also at <http://www.lang.nagoya-u.ac.jp/~matsuoka/CD-Forster-3.html#VIII>.
9. Charles Dickens, *The Life and Times of Nicholas Nickleby* (NY: Dodd, Mead, [1839]1944), p. 542.
10. Michael P. Ryan, *Knowledge Diplomacy: Global Competition and the Politics of Intellectual Property* (Washington, D.C.: Brookings Institution Press, 1998), p. 80.
11. William W. Fisher III, "The Growth of Intellectual Property: A History of the Ownership of Ideas in the United States," German Version Published in Hannes Siegrist und David Sugarman, eds. *Eigentum im Internationalen Vergleich* (Gottigen: Vandenhoeck & Ruprecht, 1999), pp. 265-91.

12. David G. Post [Temple University Law School/Cyberspace Law Institute], *Some Thoughts On The Political Economy Of Intellectual Property: A Brief Look at the International Copyright Relations of the United States* (1998), <http://www.nbr.org/regional_studies/ipr/chongqing98/post_essay.html>.
13. *Herbert v. Shanley Co.,* Nos. 427, 433, Supreme Court Of The United States, 242 U.S. 591; 37 S. Ct. 232; 61 L. Ed. 511; 1917 U.S. Lexis 2158, argued January 10, 1917, January 22, 1917.
14. Peter Orlik, "American Society of Composers, Authors and Publishers (ASCAP)," *Encyclopedia of Radio* (Chicago and London: Fitzroy Dearborn Publishers, forthcoming in 2001). <http://www.fitzroydearborn.com/chicago/radioascap.htm>.
15. Robert Allen, "Collective Invention," *Journal of Economic Behavior and Organization,* Vol. 4, No. 1 (March 1983): 2.
16. Steven W. Usselman, "Patents, Engineering Professionals, and the Pipelines of Innovation: The Internalization of Technical Discovery by Nineteenth Century American Railroads," in *Learning by Doing in Markets, Firms, and Countries,* ed. Naomi R. Lamoreaux, Daniel M. G. Raff, and Peter Temin, (Chicago: University of Chicago Press 1999), pp. 68, 70.
17. Ibid., p. 71.
18. Ibid., pp. 73-74. Tanner (Railway Co.) vs. Sayles, Supreme Court 97. S. 554 Bradley.
19. David C. Mowery and Nathan Rosenberg, *Paths of Innovation: Technological Change in 20th Century America* (Cambridge: Cambridge University Press, 1998), p. 14.
20. Ibid., pp. 18-19.
21. Charles Beard and Mary Beard, *The Rise of American Civilization,* 2 vols. in one (New York: Macmillan, 1933), pp. 112-13; and Louis M. Hacker, *The Triumph of American Capitalism: The Development of Forces in American History to the End of the Nineteenth Century* (New York: Simon and Schuster, 1940), p. 387.
22. James Boyle, *Shamans, Software, and Spleens: Law and the Construction of the Information Society* (Cambridge: Harvard University Press, 1996).
23. David C. Mowery, "The Development of Industrial Research in U.S. Manufacturing," *American Economic Review,* Vol. 80, No. 2 (May 1990): 344-9.
24. Ibid.
25. David Noble, *America By Design* (Oxford University Press, 1979); Noobar R. Danielian, *AT&T: The Story of Industrial Conquest* (New York: Vanguard Press, 1939).
26. Cited in Danielian, *AT&T,* pp. 99-100.
27. Noble, *America By Design,* p. 87; internally citing Frederick Fish, who had been both general counsel to General Electric and president of AT&T.
28. Ibid., p. 85; citing Bernard J. Stern, "The Corporations as Beneficiaries," *American Scholar,* Vol. 28 (1949): 113.
29. Noble, America By Design, p. 89.
30. Cited in Michael Polanvyi, "Patent Reform," *Review of Economic Studies,* Vol. 11, No. 2 (Summer 1944): 63.
31. *Standard Oil Co. (Ind.) v. United States,* 283 U.S. 163, 167-68 (1931).
32. Cited in Floyd Lamar Vaughan, *The United States Patent System: Legal and Economic Conflicts in American Patent History* (Westport, CT: Greenwood Press, 1972), p. 48.
33. Cited in May Wong, "High-Stakes Battle Waged Over Patents for Internet Techniques, Business Methods Law: Companies Hope for Lucrative Payoff by Laying Legal Claim to Such Commonplace Features as Clicking to Jump from One Web Site to Another," *Sacramento Bee* (July 19, 2000): C 1.
34. Fisher, "The Growth of Intellectual Property: A History of the Ownership of Ideas in the United States."
35. Jan de V. Graaf, *Theoretical Welfare Economics* (Cambridge: Cambridge University Press, 1957), p. 16.
36. Sidney G. Winter, "An Essay on the Theory of Production." in S. H. Hymans, ed., *Economics and the World Around It* (Ann Arbor: University of Michigan Press, 1982), p.76.
37. Ibid., p. 78.

38. Sidney G. Winter, "On Coase, Competence, and the Corporation," *The Nature of the Firm* in Oliver E. Williamson and Sidney G. Winter, eds. (Oxford: Oxford University Press, 1991), p. 185.
39. Kenneth J. Arrow, "The Economics of Information: An Exposition," *Empirica,* Vol. 23, No. 2 (1996): 126.
40. Ibid.
41. Richard Nelson and Sidney Winter, *An Evolutionary Theory of Economic Change* (Cambridge, MA: Belknap Press, 1982).
42. Alfred D. Jr. Chandler, *The Visible Hand: The Managerial Revolution in American Business* (Cambridge, MA: The Belknap Press, 1977), p. 294.
43. Ibid., p. 334.
44. Naomi Klein, *No Space, No Choice, No Jobs, No Logo: Taking Aim At The Brand Bullies* (NY: Picador USA, 2000).
45. Ibid.
46. Friedrich A. Hayek, "'Free' Enterprise and Competitive Order," in *Individualism and Economic Order* (Chicago: University of Chicago Press, 1948), pp. 113-14.
47. Lionel Charles Robbins, *The Economic Basis of Class Conflict and Other Essays* (London: Macmillan, 1939), p. 74.
48. Milton Friedman, *Capitalism and Freedom* (Chicago: University of Chicago Press, 1962), p. 127.
49. See William D. Nordhaus, "An Economic Theory of Technological Change (in Theory of Innovation)," *American Economic Review,* Vol. 59, No. 2 (May 1969): 18-28.
50. Edwin Mansfield, Mark Schwartz, and Samuel Wagner, "Imitation Costs and Patents: An Empirical Study," *Economic Journal,* Vol. 91, No. 364 (December 1981): 913.
51. Richard Levin, Alvin Klevorick, Richard Nelson, and Sidney Winter, "Appropriating the Returns from Industrial R&D," Cowles Foundation Working Paper, 1988, p. 913; see also Mansfield, Schwartz, and Wagner, 1981, "Imitation Costs and Patents," p. 913; and see also Nancy T. Gallini, "Patent Policy and Costly Imitation," *Rand Journal of Economics,* Vol. 23, No. 1 (Spring 1992): p. 52.
52. Richard R. Nelson, "Capitalism as an Engine of Progress," *Research Policy* (1990): pp. 193-214; reprinted in Richard R. Nelson, *The Sources of Economic Growth* (Cambridge: Harvard University Press, 1996): p. 65.
53. Levin, Klevorick, Nelson, and Winter, "Appropriating the Returns from Industrial R&D."
54. Boston Consulting Group, *The Pharmaceutical Industry into Its Second Century: From Serendipity to Strategy* (Boston Consulting Group, 2000), p. 9.
55. Amy Barrett, Ellen Licking, and John Carey, "Pharmaceuticals: Addicted To Mergers?," *Business Week* (December 6, 1999): pp. 84-88.
56. Stephen S. Hall, "Claritin and Schering-Plough: A Prescription for Profit," *New York Times Magazine* (March 11, 2001): p.
57. Gardiner Harris, "Drug Makers Pair Up to Fight Key Patent Losses," *Wall Street Journal* (May 24, 2000a): p. B 1; and David E. Rosenbaum, "The Gathering Storm Over Prescription Drugs," *New York Times* (November 14, 1999): p. D 1.
58. Harris, "Drug Makers Pair Up to Fight Key Patent Losses"; Rosenbaum, "The Gathering Storm Over Prescription Drugs."
59. Sheryl Gay Stolberg and Jeff Gerth, "Holding Down the Competition: How Companies Stall Generics and Keep Themselves Healthy." *New York Times* (July 23, 2000).
60. Ove Granstrand, *The Economics and Management of Intellectual Property: Towards Intellectual Capitalism* (Edward Elgar, 2000), pp. 39 and 53; Michael P. Ryan, *Knowledge Diplomacy: Global Competition and the Politics of Intellectual Property* (Washington, D.C.: Brookings Institution Press, 1998).
61. Fred Warshofsky, *The Patent Wars: The Battle to Own the World's Technology* (NY: Wiley, 1994), p. 8; see also Robert M. Hunt, "Patent Reform: A Mixed Blessing for the U.S. Economy?," *Business Review of the Federal Reserve Bank of Philadelphia* (November-December 1999): p. 19.

62. Paul A. David, "Intellectual Property Institutions and the Panda's Thumb: Patents, Copyrights, and Trade Secrets in Economic Theory and History," in Mitchel B. Wallerstein, Mary E. Mogee, and Robin A. Schoen, eds., *Global Dimensions of Intellectual Property Rights in Science and Technology* (Washington, D.C.: National Research Council, 1993), <http://www.nap.edu/books/0309048338/html/19.html>, p. 20.
63. Published on the company website, http://www.pfizer.com/pfizerinc/policy/intellectual-propfrm.html
64. Susan K. Sell, "Multinational Corporations as Agents of Change: The Globalization of Intellectual Property Rights," in *Private Authority and International Affairs,* ed. A. Claire Cutler, Virginia Haufler, and Tony Porter (Albany: State University of New York Press, 1999): pp. 169 and 172.
65. Anon, "The Patent Is Expiring as a Spur to Innovation." *Business Week,* Industrial/Technology edition (May 11, 1981): p. 44 C.
66. Robert M. Hunt, "Patent Reform: A Mixed Blessing for the U.S. Economy?," pp. 19-20.
67. Kevin G. Rivette and David Kline, *Rembrandts in the Attic: Unlocking the Hidden Value of Patents* (Boston: Harvard Business School Press, 2000), pp. 125-26.
68. Paula Dwyer et al., "The Battle Raging Over 'Intellectual Property,'" *Business Week* (May 22, 1989): p. 79.
69. Warshofsky, *The Patent Wars: The Battle to Own the World's Technology,* p. 111.
70. See Bernard Jr. Wysocki, "In U.S. Trade Arsenal, Brains Outgun Brawn," *Wall Street Journal* (April 10, 2000): p. A 1.
71. Richard Florida and Martin Kenney, *The Breakthrough Illusion: Corporate America's Failure to Move from Innovation to Mass Production* (NY: Basic Books, 1990), p. 237.
72. George Gilder, *Microcosm: The Quantum Revolution in Economics and Technology* (NY: Simon and Schuster, 1989).
73. Warshofsky, *The Patent Wars,* p. 9.
74. Frederick M. Scherer, *Industrial Market Structure and Economic Performance,* 2nd. ed. (Chicago: Rand McNally, 1980), pp. 155 and 449.
75. Warshofsky, *The Patent Wars,* p. 9.
76. Hunt, referring to Medtronics Inc., and Intermedics Inc. and Hybritech Inc. v. Monoclonal Antibodies Inc. (Robert M. Hunt, "You Can Patent That? Are Patents on Computer Programs and Business Methods Good for the New Economy?," *Business Review of the Federal Reserve Bank of Philadelphia* [First Quarter 2001]: pp. 5-15.)
77. Warshofsky, *The Patent Wars: The Battle to Own the World's Technology,* p. 6.
78. National Science Board, *Science and Engineering Indicators, 2000* (Washington, D.C.: National Science Federation, 2000), <http://www.nsf.gov/sbe/srs/seind00/>.
79. See also Wysocki, "In U.S. Trade Arsenal, Brains Outgun Brawn."
80. Jack Valenti (Chairman and Chief Executive Officer, Motion Picture Association), "Statement Before the Committee on Ways and Means Subcommittee on Trade, Regarding U.S.-China Trade Relations and the Possible Accession of China to the World Trade Organization (June 8, 1999)," <www.mpaa.org/jack/99/99_6_8a.htm>; based on Stephen E. Siwek and Gale Mosteller, Copyright Industries in the U.S. Economy: The 1996 Report (Economists Incorporated, 1996), <http://www.iipa.com/html/pn_executive_summary.html>.
81. Henry Nau, *National Politics and International Technologye Nuclear Reactor Development in Western Europe* (Baltimore: Johns Hopkins Press, 1974), p. 21.
82. Jim Hu, "Why the Web Can't Remain Free," News.com (May 7, 2001), <http://news.cnet.com/news/0-1014-201-5846645-0.html?tag=bt_pr>.
83. Cited in Chuck Philips, "Time Warner Tunes in New Delivery Channel," *Los Angeles Times* (July 25, 2000).
84. Anna Wilde Mathews, "Citing Napster Case, Tunesmiths Accuse Labels of Double Standard," *Wall Street Journal* (May 1, 2001): p. A1.
85. Charles C. Mann, "The Heavenly Jukebox," *Atlantic Monthly* (September 2000).

86. Michael Hoover and Lisa Stokes, "Pop Music and the Limits of Cultural Critique: Gang of Four Shrinkwraps Entertainment," *Popular Music and Society,* Vol. 22, No. 3 (Fall 1998): 26.
87. Joseph Alois Schumpeter, *Capitalism, Socialism and Democracy,* 3d. ed. (New York: Harper & Row, 1950), chapter 8.
88. See Michael A. Perelman, *The Natural Instability of Markets: Expectations, Increasing Returns and the Collapse of Markets* (NY: St. Martin's Press, 1999), chapter 5.
89. Ayn Rand, "Patents and Copyrights," *Objectivist Newsletter* (May 1964); reprinted in Ayn Rand, *Capitalism: The Unknown Ideal* (NY: New American Library, 1966): pp. 125 and 128.
90. Michael Kremer, "Patent Buyouts: A Mechanism for Encouraging Innovation," *Quarterly Journal of Economics,* Vol. 113, No. 4 (November 1998): 1137-67.
91. Cited in Paul Lewis, "The Artist's Friend Turned Enemy: A Backlash Against the Copyright," *New York Times* (January 8, 2000): p. B 9.
92. Thomas L. Friedman, *The Lexus and the Olive Tree* (New York: Farrar, Straus & Giroux, 1999), p. 373.
93. Robert Burns, "Post-Cold War Worries: Cohen to Call for Strong Military in High-Tech Visit," *Associated Press* (February 18, 1999), <http://www.abcnews.go.com/sections/us/DailyNews/cohen990218.html>.
94. Larry Light, "Why Counterfeit Goods May Kill," *Business Week* (September 2, 1996): 6.
95. Declan McCullagh, "U.S.: DVD Decoder is Terrorware," *Wired News* (May 2, 2001).
96. Fritz Machlup, "Foreword," in *The Economics of the International Patent System,* ed. Edith T. Penrose (Baltimore: Johns Hopkins University Press, 1951), p. vii.
97. www.jstor.org.
98. James Bessen and Eric Maskin, "Sequential Innovation, Patents, and Imitation," Massachusetts Institute of Technology, Department of Economics, Working Paper (January 2000).

Chapter Two

1. Boyle, *Shamans, Software, and Spleens.*
2. Mark J. Plotkin, *Tales of a Shaman's Apprentice: An Ethnobotanist Searches for New Medicines* (NY: Penguin, 1993), pp. 7-8; also see Kate Ten Kerry and Sarah A. Laird, *The Commercial Use of Biodiversity: Access to Genetic Resources and Benefit-Sharing* (London: Earthscan, 1999), pp. 40-42.
3. Mark Pendergrast, *Uncommon Grounds: The History of Coffee and How It Transformed Our World* (NY: Basic Books, 1999), p. 7.
4. Lucille Brockway, *Science and Colonial Expansion: The Role of the British Royal Botanic Gardens* (New York: Academic Press, 1979).
5. Andrew Pollack, "Biological Products Raise Genetic Ownership Issues," *New York Times* (November 26, 1999).
6. Edgar Anderson, *Plants, Man and Life* (Boston: Little, Brown & Co., 1952), pp. 132-33.
7. Jerry Uhrhammer, "Wasabi's Hot Stuff For Oregon Growers," *Portland Oregonian* (March 16, 1999).
8. Cited in William Rupert Maclaurin, *Invention and Innovation in the Radio Industry* (NY: Macmillian, 1949), p. 99.
9. Ruth Schwartz Cowan, *A Social History of American Technology* (Oxford: Oxford University Press, 1997) p. 281.
10. Michael A. Heller and Rebecca S. Eisenberg, "Can Patents Deter Innovation? The Anticommons in Biomedical Research," *Science,* Vol. 280, No. 5364 (May 1, 1998): p. 698.
11. Susan J. Douglas, *Inventing American Broadcasting, 1899-1922* (Baltimore: Johns Hopkins University Press, 1987), pp. 129-31.

12. Federal Trade Commission, *Report of the Federal Trade Commission on the Radio Industry* (Washington, D.C., USGPO, 1923), p. 14.
13. Ibid., p. 15.
14. Hugh G. J. Aitken, *The Continuous Wave: Technology and American Radio, 1900-1932* (Princeton: Princeton University Press, 1985), p. 433.
15. Robert Teitelman, *Profits of Science: The America Marriage of Business and Technology* (NY: Basic Books, 1994), p. 58.
16. Federal Trade Commission, *Report of the Federal Trade Commission on the Radio Industry,* p. 16.
17. Ibid., p. 3.
18. Danielian, *AT&T,* p. 109; internally referring to Federal Trade Commission *Report of the Federal Trade Commission on the Radio Industry,* pp. 28-29.
19. Aitken, *The Continuous Wave,* p. 260.
20. Maclaurin, *Invention and Innovation in the Radio Industry,* pp. 272-73.
21. S. G. Sturmey, *The Economic Development of Radio* (London: Duckworth, 1958). p. 223; see also Scherer, *Industrial Market Structure,* pp. 447-53.
22. Leonard S. Reich, "Research, Patents, and the Struggle to Control Radio," *Business History Review,* Vol. 11, No. 2 (Summer 1977): p. 235.
23. Timothy J. Sturgeon, "How Silicon Valley Came to Be," in Martin Kenney, ed. *Understanding Silicon Valley: The Anatomy of an Entrepreneurial Region* (Stanford: Stanford University Press, 2000): p. 29.
24. Ibid., p. 28.
25. Robert P. Merges and Richard R. Nelson, "On Limiting or Encouraging Rivalry in Technical Progress: The Effect of Patent-Scope Decisions," *Journal of Economic Behavior and Organization,* Vol. 25, No. 1 (September 1994): pp. 1-24; reprinted in Richard R. Nelson, *The Sources of Economic Growth* (Cambridge: Harvard University Press, 1996): p. 135.
26. J. M. Clark, "The Basis of War-Time Collectivism," *American Economic Review,* Vol. 7, No. 4 (December 1917): 772-90; George Bittlingmayer, "Property Rights, Progress, and the Aircraft Patent Agreement," *Journal of Law and Economics,* Vol. 31, No. 1 (April 1988): 227-48.
27. Merges and Nelson, "On Limiting or Encouraging Rivalry in Technical Progress," p. 135.
28. Bittlingmayer, "Property Rights, Progress."
29. Merges and Nelson, "On Limiting or Encouraging Rivalry in Technical Progress," pp. 136-37.
30. Ibid., p. 137.
31. Raghuram G. Rajan and Luigi Zingales, "The Firm as a Dedicated Hierarchy: A Theory of the Origins and Growth of Firms," *Quarterly Journal of Economics,* Vol. 116, No. 3 (August 2001): p. 806, internally citing Tim Jackson, *Inside Intel: Andrew Grove and the Rise of the World's Most Powerful Chip Company* (NY: Dutton Books, 1997), pp. 26-27; and Amar V. Bhide, *The Origin and the Evolution of New Businesses* (NY: Oxford University Press, 2000), p. 94.
32. Mark Walsh, "Patently Ridiculous?," *Intellectual Property Magazine* (October 1996); Walsh, "Patently Ridiculous," *Internet World Magazine* (April 1997).
33. Elisabeth Bumiller, "ASCAP Tries to Levy Campfire Royalties From Girl Scouts and Regrets It," *New York Times* (December 17, 1996); see also Anon, "Girl Scouts Don't Have To Pay Fees To Sing Songs, ASCAP Said," *Wall Street Journal* (August 27, 1996b): p. B 2.
34. Anon, "American Online Faces Lawsuit by NBA Over Data on Games," *Wall Street Journal* (August 29, 1996a): p. B 14.
35. Walsh, "Patently Ridiculous," 1997.
36. Boyle, *Shamans, Software, and Spleens,* pp. 21-24.
37. Simson Garfinkel, "A Prime Patent: Legal Rights to a Number Upset Programmers and Lawyers," *Scientific American,* Vol. 273, No. 1 (July 1995): 30.
38. Seth Shulman, "PB&J Patent Punch-up," *Technology Review* (May 2001b).

39. David James, "Trademarks: Latest Word on the Internet: It's Legal Chaos," *Business Review Weekly* (Australia), 22: 4 (February 4, 2000).
40. Seth Shulman, *Owning the Future: Staking Claims on the Knowledge Frontier* (Boston: Houghton Mifflin, 1999), p. 11.
41. Ibid., p. 69.
42. James Gleick, "Patently Absurd," *New York Times Magazine* (March 12, 2000).
43. Cited in David Voss, "'New Physics' Finds a Haven at the Patent Office," *Science* (May 21, 1999): 1252-54.
44. Warshofsky, *The Patent Wars,* p. 53.
45. Anon, "Patent Nonsense: The Knowledge Monopolies," *The Economist* (April 8, 2000a).
46. Robert P. Merges, "As Many As Six Impossible Patents Before Breakfast: Property Rights for Business Concepts and Patent System Reform," *Berkeley Technology Law Journal,* Vol. 14, No. 2 (Spring 1999), <http://www.law.berkeley.edu/journals/btlj/articles/14_2/Merges/html/text.html>.
47. Anon, "Patent Nonsense."
48. Tom Abate, "Patent Stampede as Firms Try To Lock Down New Technologies," *San Francisco Chronicle* (October 18, 1999): p. B 1.
49. Voss, "'New Physics' Finds a Haven at the Patent Office."
50. Cited in Anon, "Patent Nonsense: The Knowledge Monopolies."
51. Merges, "As Many As Six Impossible Patents Before Breakfast."
52. Brenda Sandburg, "Patent Applications Flow Freely," *Legal Times* (February 22, 1999a): 12.
53. Cited in Voss, "'New Physics' Finds a Haven at the Patent Office."
54. Sandburg, "Speed Over Substance," *Intellectual Property Magazine* (March 199b).
55. Warshofsky, *The Patent Wars,* p. 247.
56. Groucho Marx, *The Groucho Letters: Letters from and to Groucho Marx* (NY: Simon and Schuster, 1967), pp. 15-16.
57. Rivette and Kline, *Rembrandts in the Attic,* pp. 199-22.
58. Ibid.
59. See Brett Glass, "Patently Unfair? The System Created to Protect the individual Inventor May Be Hindering Innovation," *Infoworld* (October 28, 1990): pp. 56-62.
60. Merges and Nelson, "On Limiting or Encouraging Rivalry in Technical Progress," p. 134.
61. Shulman, *Owning the Future,* p. 171.
62. Warshofsky, *The Patent Wars,* p. 167.
63. Ibid.
64. Stewart Yerton, "The Sky's The Limit," *The American Lawyer* (May 1993): 64.
65. Victoria Slind-Flor, "Stalking the Submarine Patent King: Will the Enemies of Jerome Lemelson Have the Last Laugh?," *IP Worldwide* (October 1999).
66. Antonio Regalado, "Tiny TechSearch Wields Patents Against Giant-Sized Competitors," *Wall Street Journal* (March 9, 2001).
67. Ian Mount, "Would You Buy a Patent License From This Man?," *eCompany* (April 2001), <http://www.ecompany.com/articles/mag/0,1640,9575,00.html>.
68. Dean Takahashi, "Intel's Efforts to Thwart a Rival Criticized as Unethical," *Wall Street Journal* (April 19, 1999).
69. Mount, "Would You Buy a Patent License From This Man?"
70. Daniel Williams, "New Russian Firm Reinvents the Bottle," *Washington Post* (July 30, 2000): A 6.
71. Seth Shulman, "Software Patents Tangle the Web," *Technology Review* (March/April 2000).
72. See also Greg Miller and Davan Maharaj, "Will Cyber Patents Stymie Hollywood Giants?," *Los Angeles Times* (September 13, 1999).
73. Sabra Chartrand, "An Internet Entrepreneur Finds a Way for Newsstand Dealers to Profit from Subscription Sales," *New York Times* (July 26, 1999): C 8.
74. Anon, "Walker Digital Gets Patent For Online Opinion Market," *Wall Street Journal* (January 20, 1999): 9 B.

75. J. William Gurley, "Patent Here, Patent There, Patent, Patent Everywhere," *News.com* (June 14, 1999).
76. Mark Gimein, "Is The Priceline.Com Founder A Genuine Inventor—Or an Intellectual-Property Parasite?," *Salon* (August 27, 1999), <http://www.salon.com/tech/feature/1999/08/27/priceline>http://www.news.com/Perspectives/Column/0,176,341,00.html.
77. William M. Bulkeley, "Patent Application Could Evolve Into Trouble for E-Commerce," *Wall Street Journal* (28 August 28, 2000): B 1.
78. Timothy J. Mullaney and Spencer E. Ante, "Info Wars." *Business Week* (June 5, 2000): EB 107-EB 116.
79. Holman W. Jr. Jenkins, "Busting the Intellectual Property Bubble," *Wall Street Journal* (March 29, 2000): A 23.
80. Tom Webber, "Battles Over Patents Could Hurt the Web," *Wall Street Journal* (November 8, 1999): B 1.
81. Samar Farah, "The War Over Patents on the Web: Who Owns an Idea?," *Christian Science Monitor* (July 27, 2000).
82. Wong, "High-Stakes Battle Waged Over Patents."
83. Jess Bravin, "Lexis Battles an Internet Upstart Over Distributing Case Law Online," *Wall Street Journal* (May 15, 2000).
84. Cited in Ned Daly and Mike Ward, "West's Information Empire," *Multinational Monitor* (December 1994).
85. Kathryn Balint, "Public Laws Owned by the Public? Think Again, Copyright Rulings Show," *San Diego Union Tribune* (May 13, 2001).
86. John R. Wilke and James Bandler, "New Digital Camera Deals Kodak a Lesson in Microsoft's Methods: Trial Use With Windows XP Gave Microsoft an Edge, Photo Firm Says," *Wall Street Journal* (July 2, 2001): A 1.
87. Warshofsky, *The Patent Wars,* pp. 170-71; also cited in Torsten Busse, "Patents Gain Favor with Software Firms," *Infoworld* (August 26, 1991): 82.
88. Robert X. Cringely, "Notes from the Field," *InfoWorld* (August 25, 2000).
89. Luke Anderson, *Genetic Engineering, Food, and Our Environment* (White River Junction, VT: Chelsea Green, 1999), pp. 71-72.
90. Francis C. Steckel, "Cartellization of the German Chemical Industry, 1918-1925." *Journal of European Economic History,* Vol. 19, No. 2 (Fall 1990): 333.
91. Ibid.; and Johann Peter Murmann and Ralph Landau, "On the Making of Competitive Advantage: The Development of the Chemical Industries of Britain and Germany Since 1850," in *Chemicals and Long-term Economic Growth: Insights from the Chemical Industry,* ed. Ashish Arora, Ralph Landau, and Nathan Rosenberg (NY: Wiley, 1998), pp. 27-70.
92. John Gimbel, *Science, Technology, and Reparations: Exploitation and Plunder in Postwar Germany* (Stanford: Stanford University Press, 1990).
93. Lee Gomes, "A Different Tune: When Its Own Assets Are Involved, Napster Is No Fan of Sharing," *Wall Street Journal* (July 7, 2000): A 1.
94. Peter Asmus, *Reaping the Wind: How Mechanical Wizards, Visionaries, and Profiteers Helped Shape Our Energy* (Washington, D.C.: Island Press, 2001), pp. 10, 158-59.
95. Tony Paterson, "US Spy Satellites 'Raiding German Firms' Secrets,'" *Electronic Telegraph* (April 11, 1999).
96. Duncan Campbell, Development of Surveillance Technology and Risk of Abuse of Economic Information (An Appraisal of Technologies for Political Control), Part 4/4: The State of the Art in Communications Intelligence (COMINT) of Automated Processing for Intelligence Purposes of Intercepted Broadband Multi-Language Leased Or Common Carrier Systems, and its Applicability to Comint Targeting And Selection, Including Speech Recognition (Brussels: European Parliament, Directorate General for Research, Directorate. A. The STOA Programme, 1999). http://www.cyber-rights.org/interception/stoa/interception_capabilities_2000.htm.
97. Referring to Scott Shane and Tom Bowman, "America's Fortress of Spies," *Baltimore Sun* (December 3, 1995).
98. James R. Woolsey, "Why We Spy on Our Allies," *Wall Street Journal* (March 17, 2000).

99. (http://www.patents.ibm.com/details?&pn=US05937422).

Chapter Three

1. Francis Narin, Kimberly S. Hamilton, and Dominic Olivastro, "The Increasing Linkage Between U.S. Technology and Public Science," *Research Policy*, Vol. 26, No. 3 (1997): 317-30.
2. Paula E. Stephan, "The Economics of Science," *Journal of Economic Literature*, Vol. 34, No. 3 (September 1996): 1227.
3. James D. Adams, "Fundamental Stocks of Knowledge and Productivity Growth," *Journal of Political Economy*, Vol. 18, No. 4 (August 1990): 673-702.
4. Ibid.
5. Stephan, "The Economics of Science," p. 1205.
6. Renate Mayntz, "Socialist Academies of Sciences: The Enforced Orientation of Basic Research at User Needs," *Research Policy*, Vol. 27 (1998): 791.
7. Merges and Nelson, "On Limiting or Encouraging Rivalry in Technical Progress," p. 137.
8. Paul A. David, "Common Agency Contracting and the Emergence of "Open Science" Institutions." *American Economic Review*, Vol. 88, No. 2 (May 1998): 17.
9. Ibid.
10. David Levy, "The Market for Fame and Fortune," *History of Political Economy*, Vol. 20, No. 4 (Winter 1998): 615-25.
11. Stephan, "The Economics of Science," p. 1201.
12. Stephan, "The Economics of Science," pp. 1201-8; Paula E. Stephan and Sharon G. Levin, *Striking the Mother Lode in Science: The Importance of Age, Place, and Time* (New York: Oxford University Press, 1992), p. 18.
13. Robert King Merton, *Sociology of Science* (Chicago: University of Chicago Press, 1973), p. 610.
14. Robert King Merton, *On the Shoulders of Giants: A Shandean Postscript* (NY: Free Press, 1965).
15. Melvin Krantzberg, "Technology and History: 'Krantzberg's Laws,'" *Technology and Culture*, Vol. 27, No. 3 (July 1986): 549.
16. Daniel G. Dupont and Richard Lardner, "Needles in a Cold War Haystack," *Scientific American*, Vol. 275, No. 4 (November 1996): 41.
17. Teitelman, *Profits of Science*, p. 26.
18. Judith Reppy, "Military R&D and the Civilian Economy," *Bulletin of the Atomic Scientists*, Vol. 41, No. 9 (October 1985): 12.
19. Mowery and Rosenberg, *Paths of Innovation*, p. 148.
20. Alfred E. Kahn, "Fundamental Deficiencies of the American Patent Law," *American Economic Review*, Vol. 30, No. 3 (September 1940), p. 479.
21. Michael Polanvyi, "Patent Reform," *Review of Economic Studies*, Vol. 11, No. 2 (Summer 1944): 70-71.
22. Eric von Hippel, *The Sources of Innovation* (Cambridge: MIT Press, 1998).
23. See Friedrich A. Hayek, "The Use of Knowledge in Society," *American Economic Review*, Vol. 35, No. 4 (September 1945): 519-30.
24. See Richard R. Nelson, "The Link Between Science and Invention: The Case of the Transistor," in Richard R. Nelson, ed. *The Rate and Direction of Inventive Activity: Economic and Social Factors* (Princeton: Princeton University Press, 1962), pp. 549-83; reprinted in Richard R. Nelson, *The Sources of Economic Growth* (Cambridge: Harvard University Press, 1996), p. 179.
25. Ibid., p. 167.
26. Ibid., p. 164.
27. Ralph Bray, Solomon Gartenhaus, Arnold Tubis, and David Cassidy, *A History of Physics at Purdue*, http://www.physics.purdue.edu/deptinfo/ppv5/postwar.html, n.d.
28. Merges and Nelson, "On Limiting or Encouraging Rivalry in Technical Progress," p. 136.

29. Dwyer et al., "The Battle Raging Over 'Intellectual Property,'" p. 87.
30. Nelson, "The Link Between Science and Invention," p. 161.
31. Nelson, "Capitalism as an Engine of Progress," p. 62.
32. Otis Port, "Can Bell Labs' Magic Survive the AT&T Breakup?," *Business Week* (October 8, 1984): 19.
33. Michael J. L. O'Connor, *The Origins of Academic Economics in the United States* (NY: Columbia University Press, 1944).
34. James C. Williams, "Frederick E. Terman and the Rise of Silicon Valley," *International Journal of Technology Management,* Vol. 16, No. 8 (1998): 751-60.
35. Sturgeon, "How Silicon Valley Came To Be."
36. Ibid.
37. Noble, *America by Design.*
38. Thorstein Velben, *The Higher Learning in America: Memorandum on the Conduct of Universities by Business Men* (NY: B. W. Huebsch, 1918).
39. Clyde W. Barrow, *Universities and the Capitalist State* (Madison: The University of Wisconsin Press, 1990), pp. 61-71.
40. Ibid., p. 86.
41. G. Pascal Zachary, *Endless Frontier: Vannevar Bush, Engineer of the American Century* (NY: Free Press, 1997), p. 183.
42. Ibid., p. 136.
43. Ibid., p. 137.
44. Ibid., p. 223.
45. Ibid., p. 402.
46. Richard Nelson and Nathan Rosenberg, "American Universities and Technical Advance in Industry," *Research Policy,* Vol. 23, No. 3 (May 1994): 324.
47. Council on Governmental Relations, "University Technology Transfer: Questions and Answers," (November 30, 1993), <http://www.cogr.edu/qa.htm>.
48. Lee Katterman, "University Technology Offices Focus Effort on Overcoming Academic Cultural Barriers," *The Scientist* (June 12, 1995): 10-11; cited in G. Kenneth Smith, "Faculty and Graduate Student Generated Inventions: Is University Ownership a Legal Certainty?," *Virginia Journal of Law and Technology,* Vol. 1, No. 4 (Spring 1997).
49. Joshua Bloomekatz, "UC System Receives Most Patents in Nation," *The Daily Californian* (October 27, 1999): 1.
50. Paul Jacobs, "Marketplace of Ideas: Selling Patents Online," *Los Angeles Times* (October 25, 1999).
51. W. W. Powell and J. Owen-Smith, "Universities and the Market for Intellectual Property in the Life Sciences," in Burton A. Weisbrod, ed. *To Profit or Not to Profit: The Commercial Transformation of the Nonprofit Sector* (Cambridge: Cambridge University Press, 1998), p. 174.
52. Nelson and Rosenberg, "American Universities and Technical Advance."
53. Richard Arman Gregory, *Discovery, Or, The Spirit and Service of Science* (NY: Macmillan, 1916), p. 3.
54. Robert M. Berdahl, "The Public University in the Twenty-First Century," Address to National Press Club (Washington, D.C., June 2, 1999), http://www.chance.berkeley.edu/cio/chancellor/sp/press_club_address.htm.
55. Jeffrey Brainard, "Hired Guns Help Colleges Feed at the Pork Barrel," *Chronicle of Higher Education,* (October 13, 2000): A 37-A 38.
56. Karen Birchard, "U. of Oxford Sells a Share in Its Future," *Chronicle of Higher Education* (15 December 2000): A 60.
57. Anon, "Former Student In Patent Fight Leaves Prison," *New York Times* (June 14, 1996c): A 14.
58. William L. Goffe and Robert P. Parks, "The Future Information Infrastructure in Economics," *Journal of Economic Perspectives,* Vol. 11, No. 3 (Summer 1997): 79.
59. Raymond Snoddy, "Corporate Profile [of Reed Elsevier]," *Times of London* (January 4, 1999): 44.

60. See Yale Braunstein,. "Information as a Commodity: Public Policy Issues and Recent Research.," in *Information Services: Economics, Management, and Technology,* ed. Robert M. Mason and John E. Creps, Jr. (Boulder: Westview Press, 1981), p. 19.
61. Peter Applebome, "Profit Squeeze for Publishers Makes Tenure More Elusive for College Teachers," *New York Times* (November 18, 1996).
62. W. Wayt Gibbs, "Information Haves and Have-Nots," *Scientific American,* Vol. 272, No. 5 (May 1995): 12-14.
63. Andrew R. Albanese, "Persistent Suitor," *Lingua Franca,* Vol. 10, No. 9 (December 2000/ January 2001): 23-24.
64. Sheila Slaughter and Larry Leslie, *Academic Capitalism: Politics, Policies and the Entrepreneurial University* (Baltimore: John Hopkins, 1997), p. 7.
65. Mario Savio, "Speech," (December 3, 1964), <http://www.fsm-a.org/stacks/mario/ mario_speech.html>.
66. Cited in Robert Buderi, "From the Ivory Tower to the Bottom Line," *Technology Review* (July/August 2000).
67. Richard Ohmann, "Historical Reflections on Accountability," *Academe* (January-February 2000): 24-29.
68. J. Linn Allen, "City Colleges Teachers Fear Their Jobs Could Go Private," *Chicago Tribune* (February 15, 2001).
69. Jonathan N. Axelrod, "Universities Learn of Start-ups' Pitfalls," *Wall Street Journal* (August 27, 1996): C1.
70. Ronald G. Ehrenberg, *Tuition Rising: Why College Costs So Much* (Cambridge: Harvard University Press, 2000).
71. David Mehegan, "Re-engineering MIT," *Boston Globe Magazine* (May 29, 2000).
72. Howard Goodman, "University Research: Whose Work is it Anyway?," *Philadelphia Inquirer* (September 12, 1993): A1; cited in Smith, "Faculty and Graduate Student Generated Inventions."
73. Cited in Buderi, "From the Ivory Tower to the Bottom Line."
74. Amy Dockser Marcus, "Bose and Arrows: MIT Seeds Inventions But Wants a Nice Cut Of Profits They Yield," *Wall Street Journal* (July 20, 1999a): A 1.
75. Marcus, "Class Struggle: MIT Students, Lured To New Tech Firms, Get Caught in a Bind.," *Wall Street Journal* (June 24, 1999b).
76. John W. Strohm, "Between Academia and Industry," *The Tech,* Vol. 119, No. 29 (July 7, 1999), < http://www-tech.mit.edu/V119/N29/col29stroh.29c.html>.
77. Ann Grimes, "Stanford University Will Launch For-Profit Web-Search Engine," *Wall Street Journal* (May 9, 2000).
78. Karen W. Arenson, "Columbia Sets Pace In Profiting Off Research," *New York Times* (August 2, 2000): p. B 1.
79. Pual Desruisseaux, "Universities Venture Into Venture Capitalism," *Chronicle of Higher Education* (May 26, 2000): A 44-A 45.
80. Darnell Little, "Mind Melds," *Business Week* (June 5, 2000): EB 116.
81. Rebecca Zacks, "The TR University Research Scorecard," *Technology Review* (July/August 2000).
82. Sylvia Nasar, "New Breed of College All-Star: Columbia Pays Top Dollar for Economics Heavy Hitter," *New York Times* (April 8, 1998a): D 1.
83. Rebecca Weiner, "Computer Science Departments Are Depleted as More Professors Test Entrepreneurial Waters," *New York Times* (August 9, 2000).
84. John McLaren, *The Worst Mistake I Ever Made* (August 2000), <http://www.geocities.com/ jem18_99/mistake.html>; see also Christopher Shea, "No Tenure, No Peace," *Lingua Franca,* Vol. 10, No. 8 (November 2000).
85. Scott Stern, "Do Scientists Pay to Be Scientists?," National Bureau of Economic Research Working Paper No. 7410 (October 1999).
86. Lawrence C. Soley, *Leasing the Ivory Tower: The Corporate Takeover of Academia* (Boston: South End Press, 1995).
87. W. Wayt. Gibbs, "Information Haves and Have-Nots."

88. Naomi Klein, *No Space, No Choice, No Jobs, No Logo,* p. 96.
89. Paul R. Ehrlich and Anne H. Ehrlich, *Betrayal of Science and Reason: How Anti-environmental Rhetoric Threatens Our Future* (Washington, D.C.: Island Press, 1996), p. 36 ff.
90. Kevin Maguire, "University Accepts Tobacco 'Blood Money,'" *The Guardian* (December 5, 2000).
91. Eyal Press and Jennifer Washburn, "The Kept University," *Atlantic Monthly,* Vol. 285, No. 3 (March 2000): 41.
92. Anne McIlroy, "Prozac Critic Sees U of T Job Revoked," *Globe and Mail* (April 14, 2001).
93. Ibid.
94. Ibid.
95. David L. Kirp, "The New U," *The Nation* (April 17, 2000): 25.
96. Herbert I. Schiller, "Supply-Side Knowledge: Information—A Shrinking Resource," *The Nation* (December 28, 1985), p. 47; citing Liz McMillen, "Quest for Profits May Damage Basic Values of Universities, Harvard's Bok Warns," *Chronicle of Higher Education,* No. 32 (April 24, 1991): Sec. B, p. 8.
97. Ibid., p. 47; citing Anthony De Palma, "Universities' Reliance on Companies Raises Vexing Questions in Research," *New York Times* (March 17, 1991).
98. Rebecca Henderson and Iain Cockburn, "Scale, Scope, and Spillovers: The Determinants of Research Productivity in Drug Discovery," *Rand Journal of Economics,* Vol. 27, No. 3 (Spring 1996): 32-59.
99. Editorial, "Tightening Grip of Big Pharma," *The Lancet,* Vol. 357, No. 9263 (2001), p. 1141.
100. Laurie Hays, "IBM Staffers Will No Longer Send Top, Top, Top, Top-Secret Memos," *Wall Street Journal* (April 4, 1995): p. B1; cited in Julia Porter Liebeskind, "Keeping Organizational Secrets: Protective Institutional Mechanisms and Their Costs," *Industrial and Corporate Change,* Vol. 6, No. 3 (September 1997): 636.
101. Marguerite Holloway, "Outbreak Not Contained: West Nile Virus Triggers a Reevaluation of Public Health Surveillance," *Scientific American* (April 2000).
102. Christian G. Daughton and Thomas A. Ternes, "Pharmaceuticals and Personal Care Products in the Environment: Agents of Subtle Change?," *Environmental Health Perspectives,* Vol. 107, Supplement 6 (December 1999): 908, <http://ehpnet1.niehs.nih.gov/docs/1999/suppl-6/907-938daughton/abstract.html>.
103. Ibid., p. 923.
104. Ronald J. Gilson "The Legal Infrastructure of High Technology Industrial Districts: Silicon Valley, Route 128, and Covenants Not to Compete," *New York University Law Review,* Vol. 74, No. 3 (June 1999): 575-629; Alan Hyde [Rutgers University], "Wealth of Shared Information: Silicon Valley's High-Velocity Labor Market, Endogenous Economic Growth, and the Law of Trade Secrets" (1998), http://andromeda.rutgers.edu/~hyde/WEALTH.htm; and AnnaLee Saxenian, *Regional Advantage: Culture and Competition in Silicon Valley and Route 128* (Cambridge: Harvard University Press, 1994).
105. AnnaLee Saxenian, "Comment on Kenney and von Burg, 'Technology, Entrepreneurship and Path Dependence: Industrial Cluster in Silicon Valley and Route 128,'" *Industrial and Corporate Change,* Vol. 8, No. 1 (March 1999): 105-10.
106. Ian C. Ballon, "Keeping Secrets: Courts Are Becoming More Amenable to Suits Aimed at Stopping Job-Hopping Techies From Revealing Trade Secrets," *IP Magazine* (March 1998).
107. Peter Waldman, "In Silicon Valley, The Conversation Comes with a Nondisclosure Form," *Wall Street Journal* (November 3, 1999): A 1.
108. Susan Zolla-Pazner, "The Professor, the University, and Industry," *Scientific American,* Vol. 268, No. 3 (March 1994): 120.
109. Susan Zolla-Pazner, "The Professor, the University, and Industry," *Scientific American,* Vol. 268, No. 3 (March 1994): p. 120.
110. United Nations Development Programme *Globalization with a Human Face* (1999), p. 67.

111. Sylvia Nasar, *A Beautiful Mind: A Biography of John Forbes Nash, Jr.* (NY: Simon & Schuster, 1998b), p. 112.
112. W. Wayt Gibbs, "The Price of Silence" *Scientific American,* Vol. 274, No. 4 (November 1996): 15-16.
113. Stephan and Levin, *Striking the Mother Lode in Science,* p. 15; citing Lewis Wolpert and Alison Richards, *A Passion for Science* (New York: Oxford University Press, 1998), p. 106.
114. Gretchen Vogel, "Long-Suppressed Study Finally Sees Light of Day," *Science,* Vol. 676, No. 2312 (April 25, 1997): 525-26.
115. Chris Adams, "FDA Could Make Abbott Pull Synthroid, Popular Thyroid Drug, From the Market," *Wall Street Journal* (June 1, 2001a): B 1.
116. Goldie Blumenstyk, "Researchers Tell of Battling for the Right to Publish Negative Findings," *Chronicle of Higher Education,* Vol. 54, No. 31 (April 9, 1999): A 44.
117. Jennifer Washburn, "Undue Influence," *American Prospect,* Vol. 12, No. 14 (August 2001), <http://www.prospect.org/print-friendly/print/V12/14/washburn-j.html>
118. David Willman, "How a New Policy Led to Seven Deadly Drugs Medicine: Once a Wary Watchdog, the U.S. Food and Drug Administration Set Out to Become a 'Partner' of the Pharmaceutical Industry," *Los Angeles Times* (December 20, 2000).
119. Jeff Gerth and Sheryl Gay Stolberg, "When Regulators Say No: Another Part of the Battle: Keeping a Drug in the Store," *New York Times* (April 23, 2000b).
120. Richard Horton, "Lotronex and the FDA: A Fatal Erosion of Integrity," *The Lancet,* Vol. 357, No. 9268 (2001): 1544.
121. Ibid.
122. Eric G. Campbell, Joel S. Weissman, Nancyanne Causino, and David Blumenthal, "Data Withholding in Academic Medicine: Characteristics of Faculty Denied Access to Research Results and Biomaterials," *Research Policy,* Vol. 29, No. 2 (February 2000): 303-12.
123. Anon, "Is the University-Industrial Complex Out of Control?," *Nature,* Vol. 409, No. 6817 (January 11, 2001): 119.
124. Sheldon Krimsky, L. S. Rothenberg, P. Stott, and G. Kyle, "Financial Interests of Authors in Scientific Journals: A Pilot Study of 14 Publications," *Science and Engineering Ethics,* Vol. 2 (1996): 395-410.
125. Henry Thomas Stelfox, Grace Chua, Keith O'Rourke, and Allan S. Detsky, "Conflict of Interest in the Debate over Calcium-Channel Antagonists," *The New England Journal of Medicine,* Vol. 338, No. 2 (January 8, 1998): 101-6.
126. Cited in Press and Washburn, "The Kept University," p. 42.
127. Alison Bass, "Drug Companies Enrich Brown Professor," *Boston Globe* (October 4, 1999): A 1.

Chapter Four

1. Alexis de Tocqueville (1835), *Democracy in America,* Philip Bradley, ed. (New York: Vintage, 1945), pp. 462, 460.
2. Vaughan, *The United States Patent System,* p. 4.
3. See Johan Huizinga, *Homo Ludens: A Study of the Play Element in Culture* (NY: Harper and Row, 1970), p. 203.
4. Alfred Marshall (1890), "Some Aspects of Competition," in Alfred C. Pigou, ed. *Memorials of Alfred Marshall* (NY: Kelley and Millman, 1956; 1st ed., 1925), p. 281.
5. David L. Hull, *Science as Process,* (Chicago: University of Chicago Press, 1988), p. 305; cited in Stephan, "The Economics of Science," p. 1203.
6. Stern, "Do Scientists Pay to Be Scientists?"
7. Ibid.
8. Schumpeter, *Capitalism, Socialism and Democracy,* p. 132.
9. John Jewkes, David Sawyers, and Richard Stillerman, *The Sources of Invention* (London: Macmillan, 1958), p. 240 n. 1.
10. Tracy Kidder, *The Soul of a New Machine* (Boston: Little, Brown, 1981).

11. Bronwyn H. Hall and Rose Marie Ham, "The Determinants of Patenting in the U.S. Semiconductor Industry, 1980-1994," National Bureau of Economic Research Working Paper No. 7062 (March 1999).
12. Stanley M. Besen and Leo J. Raskind, "An Introduction to the Law and Economics of Intellectual Property," *Journal of Economic Perspectives,* Vol. 5, No. 1 (Winter 1991):3-27.
13. Von Hippel, *The Sources of Innovation,* p. 76 ff.
14. Catherine L. Fisk, "Removing the 'Fuel of Interest' From the 'Fire of Genius': Law and the Employee-Inventor, 1830-1930," *University of Chicago Law Review,* 65: 4 (Fall 1998): pp. 1127-98.
15. Felicity Barringer, "Freelancers Win Appeal in Copyright," *New York Times* (September 28, 1999).
16. Mark Jurkowitz, "Globe Suit Highlights Net Usage Debate Freelancers' Issue Tests Online Rights," *Boston Globe* (June 15, 2000).
17. Felicity Barringer, "Times Created a Blacklist, Writers Assert," *New York Times* (September 25, 2001).
18. Jim McKay, "Server Farms Strain Local Grids: Jurisdictions Are Facing Huge Power Demands From These Digital Warehouses," *Government Technology News* (September 29, 2001).
19. Ibid.
20. Ed White, "No Decision On GMO Case Until Fall." *Western Producer* (June 29, 2000), <http://www.producer.com/articles/20000629/news/20000629news04.html>.
21. Anon., "Nipping It In The Bud: Monsanto Isn't Just Talking Tough On Seed Piracy; It's Taking Action." *PF (Progressive Farmer) Today* (August 1998).
22. Joel Mokyr, *The Lever of Riches: Technological Creativity and Economic Progress* (NY: Oxford University Press, 1990), p. 247.
23. Scherer, *Industrial Market Structure and Economic Performance,* p. 452.
24. See Lawrence Lessig, *Code and Other Laws of Cyberspace* (NY: Basic Books, 1999), p. 46; and Jim Horner, "The Case of DAT Technology: Industrial versus Pecuniary Function," *Journal of Economic Issues,* Vol. 25, No. 2 (June 1991): 449-68.
25. Brad King, "MP3 Recorder for FM: LoFi Sound," *Wired News* (July 18, 2000), <http://www.wired.com/news/culture/0,1284,37641,00.html>.
26. Ted Bridis, "Technology Industry Aims To Render MP3 Obsolete," *Wall Street Journal* (April 12, 2001).
27. Barry Fox, "Battle Stations," *New Scientist* (October 28, 2000).
28. Pamela Samuelson, "Intellectual Property and the Digital Economy: Why the Anti-Circumvention Regulations Need to Be Revised," *Berkeley Technology Law Review,* Vol. 14, No. 2 (Spring 1999).
29. Cited in Adam Liptak, "Is Litigation The Best Way To Tame New Technology?," *New York Times* (September 2, 2000).
30. Ibid.
31. Michael A. Perelman, *Information, Social Relations, and the Economics of High Technology* (NY and London: St. Martin's and Macmillan, 1991).
32. United Nations Development Programme, *Globalization with a Human Face: United Nations Human Development Report* (NY: Oxford University Press, 1999), p. 69.
33. Wangari Maathai, "The Link between Patenting of Life Forms, Genetic Engineering and Food Insecurity," *Review of African Political Economy,* Vol. 25, No. 77 (September 1998): 527.
34. Cited in Jean-Pierre Berlan and R. C. Lewontin, "The Political Economy of Hybrid Corn," *Monthly Review,* Vol. 38, No. 3 (July-August 1986): 44.
35. Ibid., p. 45.
36. Geoffrey Lean, Volker Angres, and Louise Jury, "GM Genes 'Can Spread To People and Animals,'" *Independent* (London) (May 28, 2000).
37. Andrew Pollack, "The Green Revolution Yields to the Bottom Line," *New York Times* (May 15, 2001).
38. Ibid.

39. Perelman, *The Natural Instability of Markets.*
40. Adam Smith (1776), *The Nature and Causes of the Wealth of Nations* (Oxford: Oxford University Press, 1976), I.x.b.35, p. 128.
41. Fredrick M. Scherer, *Industry Structure, Strategy and Public Policy* (NY: Harper Collins, 1996), pp. 366-76.
42. See http://www.pathfinder.com/fortune/fortune500.
43. Marcia Angell, "The Pharmaceutical Industry—To Whom Is It Accountable?," *The New England Journal of Medicine,* Vol. 342, No. 25 (June 22, 2000).
44. Joseph DiMasi, Ronald W. Hansen, Henry G. Grabowski, and Louis Lasagna, "Costs of Innovation in the Pharmaceutical Industry," *Journal of Health Economics,* Vol. 10, No. 1 (May 1991): 107-42.
45. Jeff Gerth and Sheryl Gay Stolberg, "Medicine Merchants: Drug Companies Profit from Research Supported by Taxpayers," *New York Times* (April 23, 2000a).
46. Congressional Budget Office, "How Increased Competition from Generic Drugs Has Affected Prices and Returns in the Pharmaceutical Industry," (July 1998), chap. 3; James Love, "Call for More Reliable Costs Data on Clinical Trials," (January 13, 1997), <http://www.cptech.org/pharm/marketletter.html> http://www.cbo.gov/showdoc.cfm?index=655&sequence=0&from=0#anchor.
47. Elisabeth Rosenthal, "Research, Promotion and Profits: Spotlight Is on the Drug Industry," *New York Times* (February 21, 1993): A1.
48. Love, "Call for More Reliable Costs Data."
49. Rosenthal, "Research, Promotion and Profits."
50. Melody Petersen, "What's Black and White and Sells Medicine?," *New York Times* (August 27, 2000a).
51. Gardiner, Harris, "How Merck Survived While Others Merged—Drug Maker Relied on Inspired Research." *Wall Street Journal* (January 10, 2000c): A 1.
52. Ibid.
53. AIDS Action, "Silence = $" (1999), <media@aidsaction.org>.
54. Gardiner Harris, "Drug Firms, Stymied in the Lab, Become Marketing Machines," *Wall Street Journal* (July 5, 2000b): A 1.
55. Chris Adams, "Doctors 'Dine 'n' Dash' in Style, As Drug Firms Pick Up the Tab," *Wall Street Journal* (May 14, 2001b): A 1.
56. David D. Kirkpatrick, "Inside the Happiness Business," *New York Magazine* (May 15, 2000).
57. Adams, "Doctors 'Dine 'n' Dash' in Style."
58. Gerth and Stolberg, "Medicine Merchants."
59. Harris, "Drug Firms, Stymied in the Lab."
60. Amy Barrett, "Crunch Time in Pill Land: Behind All the Frantic Drug Mergers," *Business Week* (November 22, 1999): 52-54; Barrett et al., "Pharmaceuticals."
61. Dean Baker, "The Real Drug Crisis." *In These Times* (August 22, 1999): 19-21.
62. Ibid.
63. Marjorie Shaffer, "Survey of Pharmaceuticals," *Financial Times* (July 23, 1992): 29.
64. Victoria A. Brownworth and Bob Lederer, "The Price May Not Be Right," *POZ* (April 1997), <http://www.thebody.com/poz/backissues/4_97/policyfeature.html>.
65. Melissa B. Jacoby, Teresa A. Sullivan, and Elizabeth Warren, "Medical Problems and Bankruptcy Filings," *Norton's Bankruptcy Adviser,* Vol. 5 (2000): 1-12.
66. World Health Organization, *World Health Report 2000* (Geneva: World Health Organization, 2000), Statistical Annex, Table 1.
67. Dwyer, "The Battle Raging Over 'Intellectual Property,'" p. 78.
68. Glenn Zorpette, "All Doped Up and Going for the Gold," *Scientific American,* Vol. 282, No. 5 (May 2000): 20-22.
69. Andrew Pollack, "Two Paths to the Same Protein," *New York Times* (March 28, 2000a).
70. Andrew Pollack, "Amgen Wins Court Battle Over Drug For Anemia," *New York Times* (January 20, 2000d).

71. Andrew Pollack, "Columbia Gets Help From Alumnus on Patent Extension," *New York Times* (May 21, 2000b): A 37.
72. Boston Consulting Group, *The Pharmaceutical Industry into Its Second Century,* (2000), p. 18.
73. Bill Richards, "Baxter Beat CellPro in Court, But Some Say Dying Patients Lose," *Wall Street Journal* (August 6, 1999): A 1.
74. Gary Stix, "The Mice That Warred," *Scientific American,* 284: 6 (June 2001a): 34.
75. Ibid., p. 35.
76. Boyle, *Shamans, Software, and Spleens.*
77. http://www.cptech.org/ip/health/sa/aidsgdp1.html.
78. Karen A. Stanecki, "The AIDS Pandemic in the 21st Century: The Demographic Impact in Developing Countries," Paper at the 13th International AIDS Conference, Durban, South Africa (July 9-14, 2000).
79. Gardiner Harris, "AIDS Gaffes in Africa Come Back To Haunt Drug Industry in the U.S.," *Wall Street Journal* (April 23, 2001): A 1.
80. Donald G. Jr. McNeil, "Drug Companies and the Third World: A Case Study in Neglect," *New York Times* (May 21, 2000a).
81. Pamela Sebastian Ridge, "Business Bulletin: Aging Cats and Dogs Get More Attention From Drug," *Wall Street Journal* (November 16, 2000): A 1.
82. Joe Stephens, "The Body Hunters: As Drug Testing Spreads, Profits and Lives Hang in Balance," *Washington Post* (December 17, 2000): A01.
83. Ibid.
84. Eyal Press and Jennifer Washburn, "The Kept University," *Atlantic Monthly,* Vol. 285, No. 3 (March 2000): 48.
85. Ken Silverstein, "Millions for Viagra, Pennies for Diseases of the Poor," *The Nation* (July 19, 1999).
86. United Nations Development Programme, *Globalization with a Human Face,* p. 70.
87. McNeil, "Drug Companies and the Third World."
88. Ibid.
89. Donald G. McNeil Jr., "Study Says Combating Malaria Would Cost Little," *New York Times* (April 25, 2000b).
90. F. D. McCarthy, H. Wolf, and Y. Wu, "The Growth Costs of Malaria," National Bureau of Economic Research Working Paper No. 7541 (2000).
91. Madhusree Mukerjee, "The Berry and the Parasite," *Scientific American,* Vol. 274, No. 4 (April 1996): 22-26.
92. Mark Schoofs, "Ebola Researchers Report Progress, But a Long Road Remains Ahead," *Wall Street Journal* (November 30, 2000): B 1.
93. Pascal Fletcher, "Vaccine Gives Shot in Arm to Cuban Biotechnology Sector," *Financial Times* (London) (July 31, 1999): 4.
94. Gardiner Harris and Thomas M. Burton, "Pharmacia Is Sued After University Gets Drug Patent," *Wall Street Journal* (April 13, 2000): A 6; Patricia Grogg, "Tasty Mango Also Fights Diseases," *InterPress Service* (July 30, 2000).
95. Harris and Burton, "Pharmacia Is Sued After University Gets Drug Patent."
96. Melody Petersen, "Pushing Pills With Piles Of Money," *New York Times* (October 5, 2000b).
97. Harris and Burton, "Pharmacia Is Sued After University Gets Drug Patent."
98. Antonio Regalado, "The Great Gene Grab," *Technology Review* (September/October 2000): 53.
99. Ibid.
100. Naomi Aoki, "Patent Applications Booming In Biotech Strides in Human Genetic Code, Drive To Accrue Intellectual Capital Cited," *Boston Globe* (August 30, 2000): D 1.
101. Boston Consulting Group, *The Pharmaceutical Industry into Its Second Century,* p. 18.
102. Peter G. Gosselin and Paul Jacobs, "Patent Office Now at Heart of Gene Debate," *Los Angeles Times* (February 7, 2000a).

103. Alex Berenson and Nicholas Wade, "A Call for Sharing of Research Causes Gene Stocks to Plunge," *New York Times* (March 15, 2000): A 1.
104. King, "MP3 Recorder for FM: LoFi Sound."
105. Ibid.
106. Gosselin and Paul Jacobs "Patent Office Now at Heart of Gene Debate."
107. Anon, "New Patent Rules Pending," *Technology Review* (September/October 2000b).
108. Paul Jacobs and Peter G. Gosselin, "DNA Device's Heredity Scrutinized by U.S," *Los Angeles Times* (May 14, 2000b).
109. Leslie Roberts, "Genome Patent Fight Erupts," *Science,* Vol. 254 (October 11, 1991): 184.
110. Paul Jacobs and Peter G. Gosselin, "DNA Device's Heredity Scrutinized by U.S."
111. Berenson and Nicholas Wade, "A Call for Sharing of Research Causes Gene Stocks to Plunge."
112. Andrew Pollack, "U.S. Hopes to Stem Rush Toward Patenting of Genes," *New York Times* (June 28, 2000c).
113. Cited in Rebecca S. Eisenberg, "Re-Examining the Role of Patents in Appropriating the Value of DNA Sequences," *Emory Law Journal,* 49 (2000): 783-800.
114. Gary Stix, "Code of the Code," *Scientific American,* 284: 6 (June 2001b): 32.
115. Eisenberg, "Re-Examining the Role of Patents," p. 791.
116. Andrew Holtz, "The Role of Genetic Mutations in Breast and Ovarian Cancer," Sapient Health Network (1998),<http://nasw.org/users/holtza/SHNBRCA12.html>.
117. B. A. Koenig et al., "Genetic Testing for BRCA1 and BRCA2: Recommendations of the Stanford Program in Genomics, Ethics, and Society," *Journal of Women's Health,* Vol. 7, No. 5 (1998): 531-45.
118. Koenig et al., "Genetic Testing for BRCA1 and BRCA2."
119. Emma Ross, "British Scientists Object To Utah Company's Gene Patent Attempt," *Associated Press* (January 19, 2000).
120. Gosselin and Paul Jacobs "Patent Office Now at Heart of Gene Debate."
121. Mildred Cho, "Ethics and Legal Issues of the 21st Century," in *Preparing for the Millennium: Laboratory Medicine in the 21st Century* (Washington, D.C.: American Association for Clinical Chemistry, 1998).
122. Boston Consulting Group, *The Pharmaceutical Industry into Its Second Century,* p. 11.
123. Noe Zamel, "In Search of the Genes of Asthma on the Island of Tristan da Cunha," *Canadian Respiratory Journal,* Vol. 2, No. 1 (1995): 18-22.
124. Carolyn Abraham, "The Asthma Project: A Cautionary Tale," *Globe and Mail* (December 12, 1998): A 16.
125. Patricia Kahn, Gene Hunters Close in on Elusive Prey," *Science,* Vol. 271, No. 5254 (March 8, 1996): 1352.
126. Jeff Gulcher and Kari Stefansson, "The Icelandic Healthcare Database," *Medscape Molecular Medicine* (August 1999), http//www.medscape.com/medscape/MolecularMedicine/journal/1999/vO1 .n08/mmm6872.stef/mmm6872.stef-01 html.
127. Michael Specter, "Decoding Iceland," *New Yorker* (January 18, 1999): 40-53.
128. Gulcher and Stefansson, "The Icelandic Healthcare Database."
129. Einar Arnason, Hlynur Sigurgislason, and Eirikur Benedikz, "Genetic Homogeneity of Icelanders: Fact or Fiction?," *Nature Genetics,* Vol. 25, No. 4 (August 2000): 375.
130. Kristen Philipkoski, "Genetics Scandal Inflames Iceland," *Wired News* (March 20, 2000), <http://www.wired.com/news/politics/0,1283,35024,00.html>.
131. Specter, "Decoding Iceland," p. 53.
132. Philipkoski, "Genetics Scandal Inflames Iceland."
133. Gina Kolata, "A Special Report: Who Owns Your Genes?," *New York Times* (May 15, 2000).
134. Albert R. Hunt, "Who Will Take Their Place?," *Wall Street Journal* (May 18, 2000): A 27.
135. Newt Gingrich, "We Must Fund The Scientific Revolution," *Washington Post* (October 22, 1999): A 19.
136. Gary S. Becker, "Economic Viewpoint: Save Some of the Surplus for Medical Research," *Business Week* (April 19, 1999).

137. Kevin Murphy and Robert Topel, "The Economic Value of Medical Research" (1999), <http://www.src.uchicago.edu/users/gsb1/Rational/murphy.pdf>; later published as Kevin Murphy and Robert Topel, Exceptional Returns: The Economic Value of America's Investment in Medical Research (2000), <http://www.laskerfoundation.org/fundingfirst/papers/Funding20First.pdf>.
138. Chris Adams and Gardiner Harris, "When NIH Helps Discover Drugs, Should Taxpayers Share Wealth?," *Wall Street Journal* (June 5, 2000): B 1.
139. Gerth and Stolberg, "Medicine Merchants."
140. United States Congress, Office of Technology Assessment, *Pharmaceutical R&D: Costs, Risks and Rewards,* OTA-H-522 (Washington, DC: U.S. Government Printing Office, February 1993), p. 37.
141. Gerth and Stolberg, "Medicine Merchants."
142. United States Congress Office of Technology Assessment 1992 provided the bulk of the information for this section. (United States Congress, Office of Technology Assessment, *Federal and Private Roles in the Development and Provision of Alglucerase Therapy for Gaucher Disease,* OTA-BP-H-104 [Washington, DC: .S. Government Printing Office, October 1992].)
143. Edwin Mansfield, "Academic Research and Industrial Innovation," in Edwin Mansfield. *Innovation, Technology and the Economy: Selected Essays of Edwin Mansfield,* Vol. 2 (Aldershot, U.K.: Edward Elgar, [1991]1995), pp. 163-74.
144. Edwin Mansfield, "Academic Research and Industrial Innovation: An Update of Empirical Findings," *Research Policy,* Vol. 26, No. 7-8 (April 1998): 774.
145. Ibid., pp. 774-75.
146. Fernand Martin, "The Economic Impact of Canadian University R&D," *Research Policy,* Vol. 27, No. 7 (October 1998): 679.
147. Laurie Garrett, *Betrayal of Trust* (NY: Hyperion, 2000), p. 10.
148. Ibid., p. 200; see also p. 74.
149. Katie Hafner, "Technology Boom Too Tempting for Many Government Scientists," *New York Times* (September 19, 2000).
150. Jamie Kitman, "The Secret History of Lead," *The Nation* (March 20, 2000): 19, 26.
151. Ibid., p. 11.
152. Kitman, "The Secret History of Lead."

Chapter Five

1. See the discussion in Michael A. Perelman, *Transcending the Economy: On the Potential of Passionate Labor and the Wastes of the Market* (NY: St. Martin's Press, 2000), chapter 4.
2. See Perelman, *The Natural Instability of Markets,* chapter 7.
3. Herb Brody. "Internet@crossroads.$$$," *Technology Review,* 94: 8 (May/June 1995): pp. 30.
4. See Diane Coyle, *The Weightless World: Strategies for Managing the Digital Economy* (Cambridge: MIT Press, 1998).
5. Tom Peters, *The Circle of Innovation* (NY: Alfred A. Knopf, 1997), p. 16.
6. Alan Greenspan, "Remarks at the 80th Anniversary Awards Dinner of The Conference Board," (NY: New York, October 16, 1996).
7. Rand, "Patents and Copyrights," p. 125.
8. Ibid.
9. Gilder, *Microcosm,* p. 18.
10. Peter Huber and Mark P. Mills, "Dig More Coal—The PCs Are Coming," *Forbes* (May 31, 1999).
11. Peter Huber and Mark P. Mills, "Got a Computer? More Power to You," *Wall Street Journal* (September 7, 2000): A 26.
12. Payal Sampat, "Editorial: The New Flat Earth Society," *World-Watch Magazine* (November/December 1999); and Jonathan Koomey, *Memorandum* (LBNL-44698), Lawrence

Berkeley National Laboratories, (December 9, 1999), <http://enduse.lbl.gov/SharedData/IT/Forbescritique991209.pdf>.

13. Speech given on September 29, 2000.
14. McKay, "Server Farms Strain Local Grids."
15. Robert Bryce, "Power Struggle," *Interactive Week* (December 18, 2000), <http://www4.zdnet.com:80/intweek/stories/news/0,4164,2665339,00.html>.
16. Iddo K. Wernick, Robert Herman, Shekhar Govind, and Jesse H. Ausubel, "Materialization and Dematerialization: Measures and Trends," *Daedalus,* Vol. 125, No. 3 (Summer 1996): 171-198, <http://phe.rockefeller.edu/Daedalus/Demat/>.
17. Hays, *Behind the Silicon Curtain.*
18. John C. Dvorak, "Our Legacy: Computer Junk," *PC Magazine* (February 6, 2001): 91.
19. Paul Krugman, "The Right, the Rich, and the Facts: Deconstructing the Income Distribution Debate," *American Prospect* (Fall 1992): 23.
20. David Slawson, *The New Inflation: The Collapse of Free Markets* (Princeton: Princeton University Press, 1981), p. 51.
21. Cited in Robert J. Shiller, *Irrational Exuberance* (Princeton: Princeton University Press, 2000), p. 103.
22. Dennis C. Mueller, "Lessons From the United States' Antitrust History," *International Journal of Industrial Organization,* Vol. 14, No. 4 (June 1996): 415-16.
23. Lynn Spink, "Right to Poverty: Fear and Loathing at a Fraser Institute Right to Work Conference," *This Magazine* (Canada) 30: 4 (January/February 1997): 21.
24. Michael A. Perelman, *The End of Economics* (London: Routledge, 1996).
25. Perelman, *The Natural Instability of Markets.*
26. See Lionel Charles Robbins, *An Essay on the Nature and Significance of Economic Science,* 2d ed. (London: Macmillan, 1969), p. 16.
27. Arnold Plant, "The Economic Theory Concerning Patents for Inventions," *Economica,* ns. Vol. 1, No. 1 (February 1934): 30.
28. Arrow, "The Economics of Information," p. 125.
29. Smith, *The Nature and Causes of the Wealth of Nations,* I.iii.8, p. 363.
30. Perelman, *Information, Social Relations, and the Economics.*
31. Richard R. Nelson, "The Role of Knowledge in R&D Efficiency," *Quarterly Journal of Economics,* Vol. 97, No. 3 (August 1982): 466-67; reprinted in Richard R. Nelson, *The Sources of Economic Growth* (Cambridge: Harvard University Press, 1996), pp. 147-58.
32. Kenneth J. Arrow, "Economic Welfare and the Allocation of Resources for Invention," in *The Rate and Direction of Inventive Activity: Economic and Social Factors,* ed. Richard R. Nelson (Princeton: Princeton University Press, 1962), pp. 614-16.
33. Ibid., p. 615.
34. Merges, "The Economic Impact of Intellectual Property Rights," p. 104.
35. 35. Charles Jones and John Williams, "Measuring the Social Return to R&D," *Quarterly Journal of Economics,* Vol. 113, No. 4 (November 1998): 1119-36.
36. William J. Baumol and David Bradford, "Optimal Departures From Marginal Cost Pricing," *American Economic Review,* 60: 3 (June 1970): p. 265.
37. Ronald H. Coase, "The Lighthouse in Economics," *Journal of Law and Economics,* Vol. 17 (October 1974): pp. 357-76; reprinted in Samuel H. Baker and Catherine S. Elliott, eds. *Readings in Public Sector Economics* (Lexington, MA: D.C. Heath), pp. 256-69.
38. Hal Varian, "Markets for Public Goods," *Critical Review,* Vol. 7, No. 4 (Fall 1993): 539-57.
39. Paul A. Samuelson, "The Pure Theory of Public Expenditures," *Review of Economics and Statistics,* Vol. 36 (1954): 387-99.
40. Paul A. Samuelson, "Aspects of Public Expenditure Theories." *Review of Economics and Statistics,* Vol. 40 (November 1958): 335.
41. Jora Minasian, "Television Pricing and the Theory of Public Goods," *Journal of Law and Economics,* Vol. 7 (October 1964): 71-80.
42. Paul A. Samuelson, "Public Goods and Subscription TV," *Journal of Law and Economics,* Vol. 7 (October 1964): 83.

43. Paul A. Samuelson, "Pure Theory of Public Expenditure and Taxation," in J. Margolis and H. Guitton, eds., *Public Economics: An Analysis of Public Production and Consumption and Their Relations to the Private Sector* (London: St. Martin's Press, 1969), p. 108.
44. See Harold Demsetz, "The Private Production of Public Goods," *Journal of Law and Economics,* Vol. 13, No. 2 (October 1970): 273-306.
45. Robert Pepper [Chief, FCC Office of Plans and Policy], Letter to Senator Joseph Lieberman (September 6, 1995); cited in Ralph Nader, Testimony Before the Committee on the Budget of the U.S. House of Representatives (June 30, 1999), <http:// www.house.gov/budget/hearings/nader63099.html>.
46. Perelman, *The Natural Instability of Markets.*
47. United States Department of Agriculture, National Agricultural Statistics Service, *Agricultural Statistics 2000* (Washington, D.C.: U.S. Government Printing Office, 2000), Table 9-1, <http://www.usda.gov/nass/pubs/agr00>.
48. Granstrand, *The Economics and Management of Intellectual Property,* p. 9.
49. See http://www.interbrand.com/league_chart.html.
50. Carl E. Bartecchi, Thomas D. MacKenzie, and Robert W. Schrier, "The Global Tobacco Epidemic," *Scientific American* (May 1995): 46.
51. Cited in Bryan Burrough and John Helyar, *Barbarians at the Gate: The Fall of RJR Nabisco* (NY: Harper & Row, 1990), p. 218.
52. See http://www.interbrand.com/league_chart.html.
53. Stephen R. Bond and Jason G. Cummmins, "The Stock Market and Investment in the New Economy: Some Tangible Facts and Intangible Fictions," *Brookings Papers on Economic Activity,* Vol. 1 (2000): 65.
54. Tom Vanderbilt, *The Sneaker Book: Anatomy of an Industry and an Icon* (NY: New Press, 1998), p. 43.
55. See Richard J. Barnet and John Cavanagh, *Global Dreams: Imperial Corporations and the New World Order* (NY: Simon & Schuster, 1994), p. 328.
56. Klein, *No Space, No Choice, No Jobs, No Logo,* p. 195.
57. Ibid.

Chapter Six

1. Scherer, *Industrial Market Structure and Economic Performance,* p. 454.
2. United States Patent and Trademark Office *Annual Report, 1998* (1999); <http:// www.uspto.gov/web/offices/com/annual/1998/a98r-1.htm#Topic9>.
3. Warshofsky, *The Patent Wars,* p. 168.
4. 4. Dwyer et al., "The Battle Raging Over 'Intellectual Property,'" p. 82.
5. Andrea Gerlin, "Patent Lawyers Forgo Sure Fees on a Bet," *Wall Street Journal* (June 24, 1994): B1.
6. Mount, "Would You Buy a Patent License From This Man?"
7. Warshofsky, *The Patent Wars,* p. 254.
8. Boston Consulting Group, *The Pharmaceutical Industry into Its Second Century,* p. 12.
9. Center for Responsive Politics (1998), <http://www.opensecrets.org/lobbyists/98profiles/ 340.htm>.
10. Perelman, *Transcending the Economy,* chapter 4.
11. David P. Hamilton, "Music-Industry Group Moves to Quash Professor's Study of Antipiracy Methods," *Wall Street Journal* (May 24, 2001): A2.
12. Lawrence Lessig, "Free Code, Free Culture," O'Reilly Peer-To-Peer Conference (March 6, 2001),<http://technetcast.ddj.com/tnc_play_stream.html?stream_id=517>.
13. Seth Shulman, "Software Patents Tangle the Web," *Technology Review* (March/April 2001c).
14. William S. Comanor and Robert H. Smiley, "Monopoly and the Distribution of Wealth," *Quarterly Journal of Economics,* Vol. 89, No. 2 (May 1975): 177-94; reprinted in F. M.

Scherer, ed. *Monopoly and Competition Policy,* 2 vols. (Aldershott, Hants: Edward Elgar), p. 292.

15. Irene Powell, "The Effect of Reductions in Concentration on Income Distribution," *Review of Economics and Statistics,* Vol. 49, No. 1 (February 1987): 75-82.
16. Denise Caruso, "Technology Has Transformed the Economy and Made Some People a Lot of Money, But Is That All There Is?," *New York Times* (March 27, 2000): C 4.
17. Andrew Edgecliffe-Johnson, "Buffett Warns Against e-Hype," *Financial Times* (May 1, 2000).
18. See Seth Shulman, "IP's Bleak House," *Technology Review* (March/April 2001a).
19. Robert K. Frank, "Does Growing Inequality Harm the Middle Class?," *Eastern Economic Journal,* Vol. 26, No. 3 (Summer 2000): 263-64.
20. Deborah Reed, *California's Rising Income Inequality: Causes and Concerns* (San Francisco: Public Policy Institute of California, 1999), pp. viii-ix.
21. Mary C. Daly and Heather N. Rover, "Cyclical and Demographic Influences on the Distribution of Income in California," *Economic Review of the Federal Reserve Bank of San Francisco* (2000): 12.
22. Richard Zitrin and Carol M. Langford, "The Moral Compass: Law Firms Seem to be Stealing Signs from Baseball on Salary Issues," *American Lawyer Media* (February 4, 2000), <http://www.lawnewsnetwork.com/stories/A15267-2000Feb4.html>.
23. Emerson W. Pugh, *Memories That Shaped an Industry* (Cambridge: MIT Press, 1984), pp. 250-51; and Emerson W. Pugh, *Building IBM: Shaping an Industry and its Technology* (Cambridge: MIT Press, 1995), p. 191.
24. Anon., "A Squeeze in the Valley." *The Economist* (October 5, 2000c).
25. Ibid.
26. Evelyn Nieves, "Many in Silicon Valley Cannot Afford Housing, Even at $50,000 a Year," *New York Times* (February 20, 2000a): A 20.
27. Aaron Bernstein, "Down and Out in Silicon Valley," *Business Week* (March 27, 2000): 80.
28. Cited in Steven Greenhouse, "Janitors Struggle at the Edges of Silicon Valley's Success," *New York Times* (April 18, 2000): A18.
29. Evelyn Nieves, "In San Francisco, Renters are Humble Supplicants," *New York Times* (June 6, 2000b).
30. United States Department of Housing and Urban Development, *The State of the Cities, 2000* (2000), <http://www.huduser.org/publications/pdf/socrpt.pdf>, p. 53.
31. Nieves, "Many in Silicon Valley Cannot Afford Housing."
32. Ibid.
33. Ibid.
34. Linda Barrington, *America's Full-Time Working Poor Reap Limited Gains in the New Economy* (Washington: D. C.: Conference Board, 2000).
35. United States Department of Housing and Urban Development, *The State of the Cities,* p. 82.
36. Ibid., p. 54; citing Supervisor Gerald Connolly, who himself was cited in Michael D. Shear and Tom Jackman, "Fairfax Boom Leaves Little Room for Poor," *Washington Post* (March 14, 2000): A1.
37. Marcia Vickers, "When Wealth Is Blown Away," *Business Week* (March 26, 2001): 39-41.
38. Sheila Muto, "About 80% of Bay Area Web Firms Will Fail, San Francisco Study Finds," Wall Street Journal (March 28, 2001a).
39. Sheila Muto, "Home Prices Begin to Decline In Upscale Parts of San Francisco," *Wall Street Journal* (March 27, 2001b).
40. Tia O'Brien, "Top VC Doerr Apologizes for Helping Fuel Dot-Com Frenzy," *San Jose Mercury News* (July 16, 2001).
41. Paul A. David, "Clio and the Economics of QWERTY," *American Economic Review,* Vol. 75, No. 2 (May 1985): 332-37.
42. Brian W. Arthur, "Competing Technologies, Increasing Returns, and Lock-in by Historical Events," *Economic Journal,* Vol. 99, No. 394 (March 1989): 116-31.

43. Thorstein Veblen, *Imperial Germany and the Industrial Revolution* (New York: Macmillan & Co., 1915).
44. Ibid., p. 132.
45. See, for example, Stanley J. Liebowitz and Stephen E. Margolis, "The Fable of the Keys," *Journal of Law and Economics,* Vol. 33, No. 1 (April 1990): 1-26; and Stanley J. Liebowitz and Stephen E. Margolis, "Path Dependence and Economic Evolution," *Jobs and Capital* (Fall 1997): 36-41.
46. Holman W. Jenkins, Jr., "Don't Hate Me Because I Am Beautiful," *Wall Street Journal* (December 31, 1996), p. 23.

Concluding Thoughts

1. Timothy F. Bresnahan and M. Trajtenberg, "General Purpose Technologies: 'Engines of Growth'?" *Journal of Econometrics,* Vol. 65, No. 1 (January 1995): 83-108.

REFERENCES

Abate, Tom. 1999. "Patent Stampede as Firms Try To Lock Down New Technologies." *San Francisco Chronicle* (18 October): p. B 1.

Abraham, Carolyn. 1998. "The Asthma Project: A Cautionary Tale." *Globe and Mail* (12 December): p. A 16.

Adams, Chris. 2001a. "FDA Could Make Abbott Pull Synthroid, Popular Thyroid Drug, From the Market." *Wall Street Journal* (1 June): p. B 1.

Adams, Chris. 2001b. "Doctors 'Dine 'n' Dash' in Style, As Drug Firms Pick Up the Tab." *Wall Street Journal* (14 May): p. A 1.

Adams, Chris and Gardiner Harris. 2000. "When NIH Helps Discover Drugs, Should Taxpayers Share Wealth?" *Wall Street Journal* (5 June): p. B 1.

Adams, James D. 1990. "Fundamental Stocks of Knowledge and Productivity Growth." *Journal of Political Economy,* Vol. 18, No. 4 (August): pp. 673-702.

AIDS Action. 1999. "Silence = $." <media@aidsaction.org>.

Aitken, Hugh G. J. 1985. *The Continuous Wave: Technology and American Radio, 1900-1932* (Princeton: Princeton University Press).

Albanese, Andrew R. 2000-2001. "Persistent Suitor." *Lingua Franca,* Vol. 10, No. 9 (December 2000/January 2001): pp. 23-24.

Allen, J. Linn. 2001. "City Colleges Teachers Fear Their Jobs Could Go Private." *Chicago Tribune* (15 February).

Allen, Robert. 1983. "Collective Invention." *Journal of Economic Behavior and Organization,* Vol. 4, No. 1 (March): pp. 1-24.

Anderson, Edgar. 1952. *Plants, Man and Life* (Boston: Little, Brown & Co.).

Anderson, Luke. 1999. *Genetic Engineering, Food, and Our Environment* (White River Junction, VT: Chelsea Green).

Angell, Marcia. 2000. "The Pharmaceutical Industry—To Whom Is It Accountable?" *The New England Journal of Medicine,* Vol. 342, No. 25 (22 June 22).

Anon. 1981. "The Patent Is Expiring as a Spur to Innovation." *Business Week,* Industrial/ Technology edition (11 May): p. 44 C.

Anon. 1996a. "American Online Faces Lawsuit by NBA Over Data on Games." *Wall Street Journal* (29 August): p. B 14.

Anon. 1996b. "Girl Scouts Don't Have To Pay Fees To Sing Songs, ASCAP Said." *Wall Street Journal* (27 August): p. B 2.

Anon. 1996c. "Former Student In Patent Fight Leaves Prison." *New York Times* (14 June): p. A 14.

Anon. 1998. "Nipping It In The Bud: Monsanto Isn't Just Talking Tough On Seed Piracy; It's Taking Action." *PF (Progressive Farmer) Today* (August). <http://www.progressive-farmer.com/issue/1298/seed/default.asp>.

Anon. 1999. "Walker Digital Gets Patent For Online Opinion Market." *Wall Street Journal* (20 January): p. 9 B.

Anon. 2000a. "Patent Nonsense: The Knowledge Monopolies." *The Economist* (8 April).

Anon. 2000b. "New Patent Rules Pending." *Technology Review* (September/October).

Anon. 2000c. "A Squeeze in the Valley." *The Economist* (5 October).

Anon. 2001. "Is the University-Industrial Complex Out of Control?" *Nature,* Vol. 409, No. 6817 (11 January): p. 119.

Aoki, Naomi. 2000. "Patent Applications Booming In Biotech Strides in Human Genetic Code, Drive To Accrue Intellectual Capital Cited." *Boston Globe* (30 August): p. D 1.

Applebome, Peter. 1996. "Profit Squeeze for Publishers Makes Tenure More Elusive for College Teachers." *New York Times* (18 November).

Arenson, Karen W. 2000. "Columbia Sets Pace In Profiting Off Research." *New York Times* (2 August): p. B 1.

Arnason, Einar, Hlynur Sigurgislason and Eirikur Benedikz. 2000. "Genetic Homogeneity of Icelanders: Fact or Fiction?" *Nature Genetics,* Vol. 25, No. 4 (August): pp. 373-74.

Arrow, Kenneth J. 1962. "Economic Welfare and the Allocation of Resources for Invention." In *The Rate and Direction of Inventive Activity: Economic and Social Factors,* ed. Richard R. Nelson (Princeton: Princeton University Press): pp. 609-24.

———. 1996. "The Economics of Information: An Exposition." *Empirica,* Vol. 23, No. 2, pp. 119-28.

Arthur, W. Brian. 1989. "Competing Technologies, Increasing Returns, and Lock-in by Historical Events." *Economic Journal,* Vol. 99, No. 394 (March): pp. 116-31.

Asmus, Peter. 2001. *Reaping the Wind: How Mechanical Wizards, Visionaries, and Profiteers Helped Shape Our Energy* (Washington, D.C.: Island Press).

Axelrod, Jonathan N. 1996. "Universities Learn of Start-ups' Pitfalls." *Wall Street Journal* (27 August): p. C1.

Baker, Dean. 1999. "The Real Drug Crisis." *In These Times* (22 August): pp. 19-21.

Balint, Kathryn. 2001. "Public Laws Owned by the Public? Think Again, Copyright Rulings Show." *San Diego Union Tribune* (13 May).

Ballon, Ian C. 1998. "Keeping Secrets: Courts Are Becoming More Amenable to Suits Aimed at Stopping Job-Hopping Techies From Revealing Trade Secrets." *IP Magazine* (March).

Barnet, Richard J. and John Cavanagh. 1994. *Global Dreams: Imperial Corporations and the New World Order* (NY: Simon & Schuster).

Barrett, Amy. 1999. "Crunch Time in Pill Land: Behind All the Frantic Drug Mergers." *Business Week* (22 November): pp. 52-54.

Barrett, Amy, Ellen Licking, and John Carey. 1999. "Pharmaceuticals: Addicted To Mergers?" *Business Week* (6 December): pp. 84-88.

Barringer, Felicity. 1999. "Freelancers Win Appeal in Copyright." *New York Times* (28 September).

———. 2001. "Times Created a Blacklist, Writers Assert." *New York Times* (25 September).

Barrow, Clyde W. 1990. *Universities and the Capitalist State* (Madison: The University of Wisconsin Press).

Barringer, Felicity. 1999. "Freelancers Win Appeal in Copyright Suit." *New York Times* (28 September): p. C 1.

Barrington, Linda. 2000. *America's Full-Time Working Poor Reap Limited Gains in the New Economy* (Washington: D. C.: Conference Board).

Bartecchi, Carl E., Thomas D. MacKenzie and Robert W. Schrier. 1995. "The Global Tobacco Epidemic." *Scientific American* (May): pp. 44-51.

Bass, Alison. 1999. "Drug Companies Enrich Brown Professor." *Boston Globe* (4 October): p. A 1.

Baumol, William J. and David Bradford. 1970. "Optimal Departures From Marginal Cost Pricing." *American Economic Review,* 60: 3 (June): pp265-84.

Bazelon, David T. 1963. *The Paper Economy* (NY: Random House).

Beard, Charles and Mary. 1933. *The Rise of American Civilization.* 2 vols. in one (New York: Macmillan).

Becker, Gary S. 1999. "Economic Viewpoint: Save Some of the Surplus for Medical Research." *Business Week* (19 April).

Berdahl, Robert M. 1999. "The Public University in the Twenty-First Century." Address to National Press Club (Washington, D.C. 2 June): http://www.chance.berkeley.edu/cio/chancellor/sp/press_club_address.htm

Berenson Alex and Nicholas Wade. 2000. "A Call for Sharing of Research Causes Gene Stocks to Plunge." *New York Times* (15 March 15): p. A 1.

Berlan, Jean-Pierre and R. C. Lewontin. 1986. "The Political Economy of Hybrid Corn." *Monthly Review,* Vol. 38, No. 3 (July-August): pp. 35-47.

Bernstein, Aaron. 2000. "Down and Out in Silicon Valley." *Business Week* (27 March 27): pp. 78-92.

Besen, Stanley M. and Leo J. Raskind. 1991. "An Introduction to the Law and Economics of Intellectual Property." *Journal of Economic Perspectives,* Vol. 5, No. 1 (Winter): pp. 3-27.

Bessen, James and Eric Maskin. 2000. "Sequential Innovation, Patents, and Imitation." Massachusetts Institute of Technology, Department of Economics, Working Paper (January).

Bhide, Amar V. 2000. *The Origin and the Evolution of New Businesses* (NY: Oxford University Press).

Birchard, Karen. 2000. "U. of Oxford Sells a Share in Its Future." *Chronicle of Higher Education* (15 December): p. A 60.

Bittlingmayer, George and Thomas W. Hazlett. 2000. "DOS Kapital: Has Antitrust Action Against Microsoft Created Value in the Computer Industry?" *Journal of Financial Economics,* Vol. 55, No. 3 (March): pp. 329-59.

Bittlingmayer, George. 1988. "Property Rights, Progress, and the Aircraft Patent Agreement." *Journal of Law and Economics,* Vol. 31, No. 1 (April): pp. 227-48.

Bloomekatz, Joshua. 1999. "UC System Receives Most Patents in Nation." *The Daily Californian* (27 October): p. 1.

Blumenstyk, Goldie. 1999. "Researchers Tell of Battling for the Right to Publish Negative Findings." *Chronicle of Higher Education,* Vol. 54, No. 31 (9 April): A 44.

Bond, Patrick. 1999. "Globalization, Pharmaceutical Pricing and South African Health Policy: Managing Confrontation with U.S. Firms and Politicians Globalization, Pharmaceutical Pricing and South African Health Policy: Managing Confrontation with U.S. Firms and Politicians." *International Journal of Health Services,* No. 4, pp. 765-92.

Bond, Stephen R and Jason G. Cummmins. 2000. "The Stock Market and Investment in the New Economy: Some Tangible Facts and Intangible Fictions." *Brookings Papers on Economic Activity,* Vol. 1, pp. 61-115.

Borger, Julian. 2000. "Cuba Winning Cancer Race: Economic Isolation and a Passion for Healthcare Yield a World Lead." *Manchester Guardian* (27 July).

Boston Consulting Group. 2000. *The Pharmaceutical Industry into Its Second Century: From Serendipity to Strategy* (Boston Consulting Group).

Boyle, James. 1996. *Shamans, Software, and Spleens: Law and the Construction of the Information Society* (Cambridge: Harvard University Press).

Brainard, Jeffrey. 2000. "Hired Guns Help Colleges Feed at the Pork Barrel." *Chronicle of Higher Education,* (13 October): pp. A 37-A 38.

Braunstein, Yale. 1981. "Information as a Commodity: Public Policy Issues and Recent Research." in *Information Services: Economics, Management, and Technology,* ed. Robert M. Mason and John E. Creps, Jr. (Boulder: Westview Press).

Bravin, Jess. 2000. "Lexis Battles an Internet Upstart Over Distributing Case Law Online." *Wall Street Journal* (15 May).

Bray, Ralph, Solomon Gartenhaus, Arnold Tubis, and David Cassidy. n.d. *A History of Physics at Purdue.* <http://www.physics.purdue.edu/deptinfo/ppv5/postwar.html>

Bresnahan, Timothy F. and M. Trajtenberg. 1995. "General Purpose Technologies: `Engines of Growth'?" *Journal of Econometrics,* Vol. 65, No. 1 (January): pp. 83-108.

Bridis, Ted. 2001. "Technology Industry Aims To Render MP3 Obsolete." *Wall Street Journal* (12 April).

Brockway, Lucille. 1979. *Science and Colonial Expansion: The Role of the British Royal Botanic Gardens* (New York: Academic Press).

Brody, Herb. 1995. "Internet@crossroads.$$$." *Technology Review,* 94: 8 (May/June): pp. 24-31.

Brown, Erika, Doug Donovan, Joanne Gordon and Peter Newcomb. 1999. "Global Billionaires." *Forbes* (5 July).

Brownworth, Victoria A. and Bob Lederer. 1997. "The Price May Not Be Right." *POZ* (April). <http://www.thebody.com/poz/backissues/4_97/policyfeature.html>

Bryce, Robert. 2000. "Power Struggle." *Interactive Week* (18 December). <http://www4.zdnet.com:80/intweek/stories/news/0,4164,2665339,00.html>

Buderi, Robert. 2000. "From the Ivory Tower to the Bottom Line." *Technology Review* (July/August).

Bulkeley, William M. 2000. "Patent Application Could Evolve Into Trouble for E-Commerce." *Wall Street Journal* (28 August): p. B 1.

Bumiller, Elisabeth. 1996. "ASCAP Tries to Levy Campfire Royalties From Girl Scouts and Regrets It." *New York Times* (17 December).

Burns, Robert. 1999. "Post-Cold War Worries: Cohen to Call for Strong Military in High-Tech Visit." Associated Press (18 February). <http://www.abcnews.go.com/sections/us/DailyNews/cohen990218.html>

Burrough, Bryan and John Helyar. 1990. *Barbarians at the Gate: The Fall of RJR Nabisco* (NY: Harper & Row).

Busse, Torsten. 1991. "Patents Gain Favor with Software Firms." *Infoworld* (26 August): p. 82.

Campbell, Duncan. 1999. Development of Surveillance Technology and Risk of Abuse of Economic Information (An Appraisal of Technologies for Political Control). Part 4/4: The State of the Art in Communications Intelligence (COMINT) of Automated Processing for Intelligence Purposes of Intercepted Broadband Multi-Language Leased Or Common Carrier Systems, and its Applicability to Comint Targeting And Selection, Including Speech Recognition (Brussels: European Parliament, Directorate General for Research, Directorate. A. The STOA Programme). http://www.cyber-rights.org/interception/stoa/interception_capabilities_2000.htm.

Campbell, Eric G., Joel S. Weissman, Nancyanne Causino, and David Blumenthal. 2000. "Data Withholding in Academic Medicine: Characteristics of Faculty Denied Access to Research Results and Biomaterials." *Research Policy,* Vol. 29, No. 2 (February): pp. 303-12.

Caruso, Denise. 2000. "Technology Has Transformed the Economy and Made Some People a Lot of Money, But Is That All There Is?" *New York Times* (27 March): p. C 4.

Center for Responsive Politics. 1998. <http://www.opensecrets.org/lobbyists/98profiles/340.htm>.

Chandler, Alfred D. Jr. 1977. *The Visible Hand: The Managerial Revolution in American Business* (Cambridge, MA: The Belknap Press).

Chartrand, Sabra. 1999. "An Internet Entrepreneur Finds a Way for Newsstand Dealers to Profit from Subscription Sales." *New York Times* (26 July): p. C 8.

Cho, Mildred. 1998. "Ethics and Legal Issues of the 21st Century." in *Preparing for the Millennium: Laboratory Medicine in the 21st Century* (Washington, D.C.: American Association for Clinical Chemistry).

C.I.R. v. Wodenhouse. 1949. 337 U.S. 369. Supreme Court decision.

Clark, J. M. 1917. "The Basis of War-Time Collectivism." *American Economic Review,* Vol. 7, No. 4 (December): pp. 772-90.

Coase, Ronald H. [1974] 1990. "The Lighthouse in Economics." *Journal of Law and Economics,* Vol. 17 (October): pp. 357-76; reprinted in Samuel H. Baker and Catherine S. Elliott, eds. *Readings in Public Sector Economics.* (Lexington, MA: D.C. Heath): pp. 256-69.

Comanor, William S. and Robert H. Smiley. 1975. "Monopoly and the Distribution of Wealth." *Quarterly Journal of Economics,* Vol. 89, No. 2 (May): pp. 177-94; reprinted in F. M. Scherer, ed. *Monopoly and Competition Policy,* 2 vols. (Aldershott, Hants: Edward Elgar): 1, pp. 276-93.

Congressional Budget Office. 1998. "How Increased Competition from Generic Drugs Has Affected Prices and Returns in the Pharmaceutical Industry." (July). http://www.cbo.gov/showdoc.cfm?index=655&sequence=0&from=0#anchor

Council on Governmental Relations. 1993. "University Technology Transfer: Questions and Answers." (30 November). <http://www.cogr.edu/qa.htm>

Cowan, Ruth Schwartz. 1997. *A Social History of American Technology* (Oxford: Oxford University Press).

Coyle, Diane. 1998. *The Weightless World: Strategies for Managing the Digital Economy* (Cambridge: MIT Press).

Cringely, Robert X. 2000. "Notes from the Field." *InfoWorld* (25 August).

Daly, Mary C. and Heather N. Rover. 2000. "Cyclical and Demographic Influences on the Distribution of Income in California." *Economic Review of the Federal Reserve Bank of San Francisco,* pp. 1-13.

Daly, Ned and Mike Ward. 1994. "West's Information Empire." *Multinational Monitor* (December).

Danielian, Noobar R. 1939. *AT&T: The Story of Industrial Conquest* (New York: Vanguard Press).

Daughton, Christian G. and Thomas A. Ternes. 1999. "Pharmaceuticals and Personal Care Products in the Environment: Agents of Subtle Change?" *Environmental Health Perspectives,* Vol. 107, Supplement 6 (December): pp. 907-38. <http://ehpnet1.niehs.nih.gov/docs/1999/suppl-6/907-938daughton/abstract.html>

David, Paul A. 1993. "Intellectual Property Institutions and the Panda's Thumb: Patents, Copyrights, and Trade Secrets in Economic Theory and History." Mitchel B. Wallerstein, Mary E. Mogee, and Robin A. Schoen, eds. *Global Dimensions of Intellectual Property Rights in Science and Technology* (Washington, D.C.: National Research Council): pp. 19-62. <http://www.nap.edu/books/0309048338/html/19.html>

———. 1985. "Clio and the Economics of QWERTY." *American Economic Review,* Vol. 75, No. 2 (May): pp. 332-37.

———. 1998. "Common Agency Contracting and the Emergence of 'Open Science' Institutions." *American Economic Review,* Vol. 88, No. 2 (May): pp. 15-21.

Davoll v. Brown. 7 F. Cas. 197 (Circuit Court, D. Massachusetts 1845).

Defoe, Daniel. 1728. *A Plan of English Commerce* (London: C. Rivington; Kress Goldsmith Collection, Reel 407, No. 6594).

De Palma, Anthony. 1991. "Universities' Reliance on Companies Raises Vexing Questions in Research." *New York Times* (17 March).

Demsetz, Harold. 1970. "The Private Production of Public Goods." Journal of Law and Economics, Vol. 13, No. 2 (October): pp. 273-306.

Desruisseaux, Paul. 2000. "Universities Venture Into Venture Capitalism." *Chronicle of Higher Education* (26 May): pp. A 44-A 45.

Dickens, Charles. 1839. *The Life and Times of Nicholas Nickleby* (NY: Dodd, Mead, 1944)

———. 1841. "Letter to John Forster (24 February)." *Dickens on America and the Americans,* ed. Michael Slater (Austin: University of Texas Press, 1978); also at <http://www.lang.nagoya-u.ac.jp/~matsuoka/CD-Forster-3.html#VIII>.

DiMasi, Joseph, Ronald W. Hansen, Henry G. Grabowski, and Louis Lasagna. 1991. "Costs of Innovation in the Pharmaceutical Industry." *Journal of Health Economics,* Vol. 10, No. 1 (May): pp. 107-42.

Douglas, Susan J. 1987. *Inventing American Broadcasting, 1899-1922* (Baltimore: Johns Hopkins University Press).

Dupont, Daniel G. and Richard Lardner. 1996. "Needles in a Cold War Haystack." *Scientific American,* Vol. 275, No. 4 (November): pp. 41 and 44.

Dvorak, John C. 2001. "Our Legacy: Computer Junk." *PC Magazine* (6 February): p. 91.

Dwyer, Paula et al. 1989. "The Battle Raging Over 'Intellectual Property'." *Business Week* (22 May): pp. 78-89.

Edgecliffe-Johnson, Andrew. 2000. "Buffett Warns Against e-Hype." *Financial Times* (1 May).

Editorial. 2001. "Tightening Grip of Big Pharma." Lancet, Vol. 357, No. 9263, p. 1141.

Ehrenberg, Ronald G. 2000. *Tuition Rising: Why College Costs So Much* (Cambridge: Harvard University Press).

Ehrlich, Paul R. and Anne H. Ehrlich. 1996. *Betrayal of Science and Reason: How Anti-environmental Rhetoric Threatens Our Future* (Washington, D.C.: Island Press).

Eisenberg, Rebecca S. 2000. "Re-Examining the Role of Patents in Appropriating the Value of DNA Sequences." *Emory Law Journal,* 49, pp. 783-800.

Farah, Samar. 2000. "The War Over Patents on the Web: Who Owns an Idea?" *Christian Science Monitor* (27 July).

Federal Trade Commission. 1923. *Report of the Federal Trade Commission on the Radio Industry* (Washington, D.C., USGPO).

Fisher, William W., III. 1999. "The Growth of Intellectual Property: A History of the Ownership of Ideas in the United States." German Version Published in Hannes Siegrist und David Sugarman, eds. *Eigentum im Internationalen Vergleich* (Gottigen: Vandenhoeck & Ruprecht): pp. 265-91.

Fisk, Catherine L. 1998. "Removing the 'Fuel of Interest' From the 'Fire of Genius': Law and the Employee-Inventor, 1830-1930." *University of Chicago Law Review*, 65: 4 (Fall): pp. 1127-98.

Fletcher, Pascal. 1999. "Vaccine Gives Shot in Arm to Cuban Biotechnology Sector." *Financial Times* (London) (31 July): p. 4.

Florida, Richard and Martin Kenney. 1990. *The Breakthrough Illusion: Corporate America's Failure to Move from Innovation to Mass Production* (NY: Basic Books).

Fox, Barry. 2000. "Battle Stations." *New Scientist* (28 October).

Frank, Robert K. 2000. "Does Growing Inequality Harm the Middle Class?" *Eastern Economic Journal*, Vol. 26, No. 3 (Summer): pp. 253-64.

Friedman, Milton. 1962. *Capitalism and Freedom* (Chicago: University of Chicago Press).

Friedman, Thomas L. 1999. *The Lexus and the Olive Tree* (Farrar Straus & Giroux).

Gallini, Nancy T. 1992. "Patent Policy and Costly Imitation." *Rand Journal of Economics*, Vol. 23, No. 1 (Spring): pp. 52-63.

Garfinkel, Simson. 1995. "A Prime Patent: Legal Rights to a Number Upset Programmers and Lawyers." *Scientific American*, Vol. 273, No. 1 (July): p. 30.

Garrett, Laurie. 2000. *Betrayal of Trust* (NY: Hyperion).

Gerlin, Andrea. 1994. "Patent Lawyers Forgo Sure Fees on a Bet." *Wall Street Journal* (24 June): p. B1.

Gerth, Jeff and Sheryl Gay Stolberg. 2000a. "Medicine Merchants: Drug Companies Profit from Research Supported by Taxpayers." *New York Times* (23 April).

———. 2000b. "When Regulators Say No: Another Part of the Battle: Keeping a Drug in the Store." *New York Times* (23 April).

Gibbs, W. Wayt. 1995. "Information Haves and Have-Nots." *Scientific American*, Vol. 272, No. 5 (May): pp. 12-14.

———. 1996. "The Price of Silence." *Scientific American*, Vol. 274, No. 4 (November): pp. 15-16.

Gilder, George. 1989. *Microcosm: The Quantum Revolution in Economics and Technology* (NY: Simon and Schuster).

Gilson, Ronald J. 1999. "The Legal Infrastructure of High Technology Industrial Districts: Silicon Valley, Route 128, and Covenants Not to Compete." New York University Law Review, Vol. 74, No. 3 (June): pp. 575-629.

Gimbel, John. 1990. *Science, Technology, and Reparations: Exploitation and Plunder in Postwar Germany* (Stanford: Stanford University Press).

Gimein, Mark. 1999. "Is The Priceline.Com Founder A Genuine Inventor—Or an Intellectual-Property Parasite?" *Salon* (27 August). <http://www.salon.com/tech/feature/1999/08/27/priceline>http://www.news.com/Perspectives/Column/0,176,341,00.html

Gingrich, Newt. 1999. "We Must Fund The Scientific Revolution." *Washington Post* (22 October): p. A 19.

Glass, Brett. 1990. "Patently Unfair? The System Created to Protect the individual Inventor May Be Hindering Innovation." *Infoworld* (28 October): pp. 56-62.

Gleick, James. 2000. "Patently Absurd." *New York Times Magazine* (12 March).

Goffe, William L. and Robert P. Parks. 1997. "The Future Information Infrastructure in Economics." *Journal of Economic Perspectives*, Vol. 11, No. 3 (Summer): pp. 75-94.

Gomes, Lee. 2000. "A Different Tune: When Its Own Assets Are Involved, Napster Is No Fan of Sharing." *Wall Street Journal* (7 July): p. A 1.

Goodman, Howard. 1993. "University Research: Whose Work is it Anyway?" *Philadelphia Inquirer* (September 12): p. A1.

Gosselin, Peter G. and Paul Jacobs. 2000a. "Patent Office Now at Heart of Gene Debate." *Los Angeles Times* (7 February).

———. 2000b. "DNA Device's Heredity Scrutinized by U.S." *Los Angeles Times* (14 May).

Graaf, Jan de V. 1957. *Theoretical Welfare Economics* (Cambridge: Cambridge University Press).

Granstrand, Ove. 2000. *The Economics and Management of Intellectual Property: Towards Intellectual Capitalism* (Edward Elgar).

Greenhouse, Steven. 2000. "Janitors Struggle at the Edges of Silicon Valley's Success." *New York Times* (18 April): p. A18.

Greenspan, Alan. 1996. "Remarks at the 80th Anniversary Awards Dinner of The Conference Board." (NY: New York, 16 October).

Gregory, Richard Arman. 1916. *Discovery, Or, The Spirit and Service of Science* (NY: Macmillan).

Grimes, Ann. 2000. "Stanford University Will Launch For-Profit Web-Search Engine." *Wall Street Journal* (9 May).

Grogg, Patricia. 2000. "Tasty Mango Also Fights Diseases." *InterPress Service* (30 July).

Gulcher, Jeff and Kari Stefansson. 1999. "The Icelandic Healthcare Database." *Medscape Molecular Medicine* (August) http//www.medscape.com/medscape/MolecularMedicine/journal/1999/vO1 .n08/mmm6872.stef/mmm6872.stef-01 html.

Gurley, J. William. 1999. "Patent Here, Patent There, Patent, Patent Everywhere." *News.com* (14 June).

Hacker, Louis, M. 1940. *The Triumph of American Capitalism: The Development of Forces in American History to the End of the Nineteenth Century* (New York: Simon and Schuster).

Hafner, Katie. 2000. "Technology Boom Too Tempting for Many Government Scientists." *New York Times* (September 19).

Hall, Bronwyn H. and Rose Marie Ham. 1999. "The Determinants of Patenting in the U.S. Semiconductor Industry, 1980-1994." National Bureau of Economic Research Working Paper No. 7062 (March 1999).

Hall, Stephen S. 2001. "Claritin and Schering-Plough: A Prescription for Profit." *New York Times Magazine* (11 March).

Hamilton, David P. 2001. "Music-Industry Group Moves to Quash Professor's Study of Antipiracy Methods." *Wall Street Journal* (24 May): p. A2.

Harris, Gardiner. 2000a. "Drug Makers Pair Up to Fight Key Patent Losses." *Wall Street Journal* (24 May): p. B 1.

———. 2000b. "Drug Firms, Stymied in the Lab, Become Marketing Machines." *Wall Street Journal* (5 July): p. A 1.

———. 2000c. "How Merck Survived While Others Merged—Drug Maker Relied on Inspired Research." *Wall Street Journal* (10 January): p. A 1.

———. 2001. "AIDS Gaffes in Africa Come Back To Haunt Drug Industry in the U.S." *Wall Street Journal* (23 April): p. A 1.

Harris, Gardiner and Thomas M. Burton. 2000. "Pharmacia Is Sued After University Gets Drug Patent." *Wall Street Journal* (13 April): A 6.

Hayek, Friedrich A. 1945. "The Use of Knowledge in Society." *American Economic Review,* Vol. 35, No. 4 (September): pp. 519-30.

———. 1948. "'Free' Enterprise and Competitive Order." in *Individualism and Economic Order* (Chicago: University of Chicago Press): pp. 107-18.

Hays, Dennis. 1989. *Behind the Silicon Curtain: The Seduction of Work in a Lonely Age* (Boston: South End Press).

Hays, Laurie. 1995. "IBM Staffers Will No Longer Send Top, Top, Top, Top-Secret Memos." *Wall Street Journal* (4 April): p. B1.

Heller. Michael. 1998. "The Tragedy of the Anticommons: Property in the Transition From Marx to Markets." *Harvard Law Review,* Vol. 111, No. 3, pp. 621-89.

Heller, Michael A. and Rebecca S. Eisenberg. 1998. "Can Patents Deter Innovation? The Anticommons in Biomedical Research." *Science,* Vol. 280, No. 5364 (1 May 1): p. 698-701.

Henderson, Rebecca and Iain Cockburn. 1996. "Scale, Scope, and Spillovers: The Determinants of Research Productivity in Drug Discovery." *Rand Journal of Economics,* Vol. 27, No. 3 (Spring): pp. 32-59.

Herbert v. Shanley Co., Nos. 427, 433, 242 U.S. 591; 37 S. Ct. 232; 61 L. Ed. 511; 1917 U.S. Lexis 2158. (Supreme Court Of The United States, January 22, 1917). Argued January 10, 1917.

Holloway, Marguerite. 2000. "Outbreak Not Contained: West Nile Virus Triggers a Reevaluation of Public Health Surveillance." *Scientific American* (April).

Holtz, Andrew. 1998. "The Role of Genetic Mutations in Breast and Ovarian Cancer." Sapient Health Network <http://nasw.org/users/holtza/SHNBRCA12.html>

Hoover, Michael and Lisa Stokes. 1998. "Pop Music and the Limits of Cultural Critique: Gang of Four Shrinkwraps Entertainment." *Popular Music and Society,* Vol. 22, No. 3 (Fall): pp. 21-38.

Horner, Jim. 1991. "The Case of DAT Technology: Industrial versus Pecuniary Function." *Journal of Economic Issues,* Vol. 25, No. 2 (June): pp. 449-68.

Horton, Richard. 2001. "Lotronex and the FDA: A Fatal Erosion of Integrity." *The Lancet,* Vol. 357, No. 9268, p. 1544.

Hu, Jim. 2001. "Why the Web Can't Remain Free." News.com (7 May). <http://news.cnet.com/news/0-1014-201-5846645-0.html?tag=bt_pr>

Huber, Peter and Mark P. Mills. 1999. "Dig More Coal—The PCs Are Coming." *Forbes* (31 May).

———. 2000. "Got a Computer? More Power to You." *Wall Street Journal* (7 September): p. A 26.

Huizinga, Johan. 1970. *Homo Ludens: A Study of the Play Element in Culture* (NY: Harper and Row).

Hull, David L. 1988. *Science as Process.* (Chicago: University of Chicago Press).

Hunt, Albert R. 2000. "Who Will Take Their Place?" *Wall Street Journal* (18 May) p. A 27.

Hunt, Robert M. 1999. "Patent Reform: A Mixed Blessing for the U.S. Economy?" *Business Review of the Federal Reserve Bank of Philadelphia* (November-December): pp. 15-29.

———. 2001. "You Can Patent That? Are Patents on Computer Programs and Business Methods Good for the New Economy?" *Business Review of the Federal Reserve Bank of Philadelphia* (First Quarter): pp. 5-15.

Hyde, Alan [Rutgers University]. 1998. "Wealth of Shared Information: Silicon Valley's High-Velocity Labor Market, Endogenous Economic Growth, and the Law of Trade Secrets." http://andromeda.rutgers.edu/~hyde/WEALTH.htm

Jackson, Tim. 1997. *Inside Intel: Andrew Grove and the Rise of the World's Most Powerful Chip Company* (NY: Dutton Books).

Jacobs, Paul. 1999. "Marketplace of Ideas: Selling Patents Online." *Los Angeles Times* (25 October).

Jacobs, Paul and Peter G. Gosselin. 2000. "'Robber Barons of the Genetic Age': Experts Fret Over Effect of Gene Patents on Research." *Los Angeles Times* (28 February).

Jacoby, Melissa B., Teresa A. Sullivan, and Elizabeth Warren. 2000. "Medical Problems and Bankruptcy Filings." *Norton's Bankruptcy Adviser,* Vol. 5, pp. 1-12.

James, David. 2000. "Trademarks: Latest Word on the Internet: It's Legal Chaos." *Business Review Weekly* (Australia), 22: 4 (4 February).

Jenkins, Holman W. Jr. 1996. "Don't Hate Me Because I Am Beautiful." *Wall Street Journal.* (December 31): p. 23.

———. 2000. "Busting the Intellectual Property Bubble." *Wall Street Journal* (29 March): A 23.

Jewkes, John, David Sawyers and Richard Stillerman. 1958. *The Sources of Invention* (London: Macmillan).

Jones, Charles and John Williams. 1998. "Measuring the Social Return to R&D." *Quarterly Journal of Economics,* Vol. 113, No. 4 (November): pp. 1119-36.

Jurkowitz, Mark. 2000. "Globe Suit Highlights Net Usage Debate Freelancers' Issue Tests Online Rights." *Boston Globe* (15 June).

Kahn, Alfred E. 1940. "Fundamental Deficiencies of the American Patent Law." *American Economic Review,* Vol. 30, No. 3 (September), p. 475-91.

Kahn, Patricia. 1996. "Gene Hunters Close in on Elusive Prey." *Science,* Vol. 271, No. 5254 (8 March): pp. 1352-4.

Katterman, Lee. 1995. "University Technology Offices Focus Effort on Overcoming Academic Cultural Barriers." *The Scientist* (June 12): pp. 10-11.

Kidder, Tracy. 1981. *The Soul of a New Machine* (Boston: Little, Brown).

King, Brad. 2000. "MP3 Recorder for FM: LoFi Sound." *Wired News* (18 July). <http://www.wired.com/news/culture/0,1284,37641,00.html>.

King, Ralph T., Jr. 2000. "Assembly-Line Sequencing Lets Firm Beat a Path to the U.S. Patent Office." *Wall Street Journal* (10 February): p. A 1.

Kirkpatrick, David D. 2000. "Inside the Happiness Business." *New York Magazine* (15 May).

Kirp, David L. 2000. "The New U." *The Nation* (17 April): pp. 25-29.

Kitman, Jamie. 2000. "The Secret History of Lead." *The Nation* (20 March): pp. 11-44.

Klein, Naomi. 2000. *No Space, No Choice, No Jobs, No Logo: Taking Aim At The Brand Bullies* (NY: Picador USA).

Koenig, B. A. et al. 1998. "Genetic Testing for BRCA1 and BRCA2: Recommendations of the Stanford Program in Genomics, Ethics, and Society." *Journal of Women's Health,* Vol. 7, No. 5, pp. 531-45.

Kolata, Gina. 2000. "A Special Report: Who Owns Your Genes?" *New York Times* (15 May).

Koomey, Jonathan. 1999. *Memorandum* (LBNL-44698) (9 December) Lawrence Berkeley National Laboratories. <http://enduse.lbl.gov/SharedData/IT/Forbescritique991209.pdf>

Krantzberg, Melvin. 1986. "Technology and History: 'Krantzberg's Laws.'" *Technology and Culture,* Vol. 27, No. 3 (July): pp. 544-60.

Kremer, Michael. 1998. "Patent Buyouts: A Mechanism for Encouraging Innovation." *Quarterly Journal of Economics,* Vol. 113, No. 4 (November): pp. 1137-67.

Krimsky, Sheldon, L. S. Rothenberg, P. Stott and G. Kyle. 1996. "Financial Interests of Authors in Scientific Journals: A Pilot Study of 14 Publications." *Science and Engineering Ethics,* Vol. 2, pp. 395-410.

Krugman, Paul. 1992. "The Right, the Rich, and the Facts: Deconstructing the Income Distribution Debate." *American Prospect* (Fall): pp. 19-31.

Lean, Geoffrey, Volker Angres and Louise Jury. 2000. "GM Genes 'Can Spread To People and Animals'." *Independent* (London) (28 May).

Lessig, Lawrence. 1999. *Code and Other Laws of Cyberspace* (NY: Basic Books).

———. 2001. "Free Code, Free Culture." O'Reilly Peer-To-Peer Conference (6 March).<http://technetcast.ddj.com/tnc_play_stream.html?stream_id=517>

Levin, Richard, Alvin Klevorick, Richard Nelson, and Sidney Winter. 1988. "Appropriating the Returns from Industrial R&D." Cowles Foundation Working Paper.

Levy, David. 1988. "The Market for Fame and Fortune." *History of Political Economy,* Vol. 20, No. 4 (Winter): pp. 615-25.

Lewis, Paul. 2000. "The Artist's Friend Turned Enemy: A Backlash Against the Copyright." *New York Times* (8 January): p. B 9.

Liebeskind, Julia Porter. 1997. "Keeping Organizational Secrets: Protective Institutional Mechanisms and Their Costs." *Industrial and Corporate Change,* Vol. 6, No. 3 (September): pp. 623-63.

Liebowitz, Stanley J. and Stephen E. Margolis. 1990. "The Fable of the Keys." *Journal of Law and Economics,* Vol. 33, No. 1 (April): pp. 1-26.

———. 1997. "Path Dependence and Economic Evolution." *Jobs and Capital* (Fall): pp. 36-41.

Light, Larry. 1996. "Why Counterfeit Goods May Kill." *Business Week* (2 September): p. 6.

Liptak, Adam. 2000. "Is Litigation The Best Way To Tame New Technology?" *New York Times* (2 September).

Little, Darnell. 2000. "Mind Melds." *Business Week* (5 June): p. EB 116.

Love, James. 1997. "Call for More Reliable Costs Data on Clinical Trials." (13 January). <http://www.cptech.org/pharm/marketletter.html>

Maathai, Wangari. 1998. "The Link between Patenting of Life Forms, Genetic Engineering and Food Insecurity." *Review of African Political Economy,* Vol. 25, No. 77 (September): pp. 526-28.

Machlup, Fritz and Edith Penrose. 1950. "The Patent Controversy in the Nineteenth Century." *Journal of Economic History,* Vol. 10, No. 1 (May): pp. 1-29.

Machlup, Fritz. 1951. "Foreword." Edith T. Penrose. *The Economics of the International Patent System* (Baltimore: Johns Hopkins University Press).

MacKay, Judge W. Andrew. 2001. Monsanto Canada Inc. and Monsanto Company Vs. Percy Schmeiser and Schmeiser Enterprises Ltd. 2001 FCT 256.

Maclaurin, William Rupert. 1949. *Invention and Innovation in the Radio Industry* (NY: Macmillian).

Maguire, Kevin. 2000. "University Accepts Tobacco 'Blood Money'." *The Guardian* (5 December).

Mandel, Michael J. 2001. "How the Super-Rich Lucked Out Twice: New Data Show The Top Earners Are Already Enjoying Lower Rates." *Business Week* (14 May): p. 52.

Mann, Charles C. 2000. "The Heavenly Jukebox." *Atlantic Monthly* (September).

Mansfield, Edwin. 1991. "Academic Research and Industrial Innovation." in Edwin Mansfield. *Innovation, Technology and the Economy: Selected Essays of Edwin Mansfield,* Vol. 2 (Aldershot, U.K.: Edward Elgar, 1995): pp. 163-74.

———. 1998. "Academic Research and Industrial Innovation: An Update of Empirical Findings." *Research Policy,* Vol. 26, No. 7-8 (April): pp. 773-76.

Mansfield, Edwin, Mark Schwartz and Samuel Wagner. 1981. "Imitation Costs and Patents: An Empirical Study." *Economic Journal,* Vol. 91, No. 364 (December): pp.: 907-18.

Martin, Fernand. 1998. "The Economic Impact of Canadian University R&D." *Research Policy,* Vol. 27, No. 7 (October): pp. 677-687.

Marcus, Amy Dockser. 1999a. "Bose and Arrows: MIT Seeds Inventions But Wants a Nice Cut Of Profits They Yield." *Wall Street Journal* (20 July): p. A 1.

———. 1999b. "Class Struggle: MIT Students, Lured To New Tech Firms, Get Caught in a Bind." *Wall Street Journal* (24 June).

Marshall, Alfred. 1890. "Some Aspects of Competition." in Alfred C. Pigou, ed. *Memorials of Alfred Marshall* (NY: Kelley and Millman, 1956; 1st ed., 1925): pp. 256-91.

Mathews, Anna Wilde. 2001. "Citing Napster Case, Tunesmiths Accuse Labels of Double Standard." *Wall Street Journal* (1 May): p. A1.

Marx, Groucho. 1967. *The Groucho Letters: Letters from and to Groucho Marx* (NY: Simon and Schuster).

Marx, Karl. 1977. *Capital.* (NY: Vintage).

Mayntz, Renate. 1998. "Socialist Academies of Sciences: The Enforced Orientation of Basic Research at User Needs." *Research Policy,* Vol. 27: pp. 781-91.

McCarthy, F. D., H. Wolf, and Y. Wu. 2000. "The Growth Costs of Malaria." National Bureau of Economic Research Working Paper No. 7541.

McCullagh, Declan. 2001. "U.S.: DVD Decoder is Terrorware." *Wired News* (2 May).

McIlroy, Anne. 2001. "Prozac Critic Sees U of T Job Revoked." *Globe and Mail* (14 April).

McKay, Jim. 2001. "Server Farms Strain Local Grids: Jurisdictions Are Facing Huge Power Demands From These Digital Warehouses." *Government Technology News* (29 September). <http://www.govtech.net/news/features/feature_sept_29.phtml>

McLaren, John. 2000. *The Worst Mistake I Ever Made* (August) <http://www.geocities.com/jem18_99/mistake.html>

McMillen, Liz. 1991. "Quest for Profits May Damage Basic Values of Universities, Harvard's Bok Warns." *Chronicle of Higher Education,* No. 32 (24 April): Sec. B, p. 8.

McNeil, Donald G. Jr. 2000b. "Study Says Combating Malaria Would Cost Little." *New York Times* (25 April).

McNeil, Donald G. Jr. 2000a. "Drug Companies and the Third World: A Case Study in Neglect." *New York Times* (21 May).

Mehegan, David. 2000. "Re-engineering MIT." *Boston Globe Magazine* (29 May).

Merges, Robert P. 1995. "The Economic Impact of Intellectual Property Rights: An Overview and Guide." *Journal of Cultural Economics,* Vol. 19, No. 2, pp. 103-17.

———. 1999. "As Many As Six Impossible Patents Before Breakfast: Property Rights for Business Concepts and Patent System Reform." *Berkeley Technology Law Journal,* Vol. 14, No. 2

(Spring). <http://www.law.berkeley.edu/journals/btlj/articles/14_2/Merges/html/text.html>

Merges, Robert P. and Richard R. Nelson. 1994. "On Limiting or Encouraging Rivalry in Technical Progress: The Effect of Patent-Scope Decisions." *Journal of Economic Behavior and Organization,* Vol. 25, No. 1 (September): pp. 1-24; reprinted in Richard R. Nelson. 1996. *The Sources of Economic Growth* (Cambridge: Harvard University Press): pp. 120-44.

Merton, Robert King. 1965. *On the Shoulders of Giants: A Shandean Postscript* (NY: Free Press).

———. 1973. *Sociology of Science* (Chicago: University of Chicago Press).

Miller, Greg and Davan Maharaj. 1999. "Will Cyber Patents Stymie Hollywood Giants?" *Los Angeles Times* (13 September).

Minasian, Jora. 1964. "Television Pricing and the Theory of Public Goods." *Journal of Law and Economics,* Vol. 7 (October): pp. 71-80.

Mitchell v. Tilghman. 86 U.S. 287.

Mokyr, Joel. 1990. *The Lever of Riches: Technological Creativity and Economic Progress* (NY: Oxford University Press).

Mount, Ian. 2001. "Would You Buy a Patent License From This Man?" *eCompany* (April). <http://www.ecompany.com/articles/mag/0,1640,9575,00.html>

Mowery, David C. 1990. "The Development of Industrial Research in U.S. Manufacturing." *American Economic Review,* Vol. 80, No. 2 (May): pp. 344-9.

Mowery, David C. and Nathan Rosenberg. 1998. *Paths of Innovation: Technological Change in 20th Century America* (Cambridge: Cambridge University Press).

Mueller, Dennis C. 1996. "Lessons From the United States' Antitrust History." *International Journal of Industrial Organization,* Vol. 14, No. 4 (June): pp. 415-46.

Mukerjee, Madhusree. 1996. "The Berry and the Parasite." *Scientific American,* Vol. 274, No. 4 (April): pp. 22-26.

Mullaney, Timothy J. and Spencer E. Ante. 2000. "Info Wars." *Business Week* (5 June): pp. EB 107-EB 116.

Murmann, Johann Peter and Ralph Landau. 1998. "On the Making of Competitive Advantage: The Development of the Chemical Industries of Britain and Germany Since 1850." in *Chemicals and Long-term Economic Growth: Insights from the Chemical Industry,* eds. Ashish Arora, Ralph Landau, and Nathan Rosenberg (NY: Wiley): pp. 27-70.

Murphy, Kevin and Robert Topel. 1999. "The Economic Value of Medical Research." <http://www.src.uchicago.edu/users/gsb1/Rational/murphy.pdf>; Later published as Murphy, Kevin and Robert Topel. 2000. Exceptional Returns: The Economic Value of America's Investment in Medical Research. <http://www.laskerfoundation.org/fundingfirst/papers/Funding20First.pdf>

Muto, Sheila. 2001a. "About 80% of Bay Area Web Firms Will Fail, San Francisco Study Finds." Wall Street Journal (28 March).

———. 2001b. "Home Prices Begin to Decline In Upscale Parts of San Francisco." *Wall Street Journal* (27 March).

Nader, Ralph. 1999. Testimony Before the Committee on the Budget of the U.S. House of Representatives (30 June). <http://www.house.gov/budget/hearings/nader63099.html>

Narin, Francis, Kimberly S. Hamilton and Dominic Olivastro. 1997. "The Increasing Linkage Between U.S. Technology and Public Science." *Research Policy,* Vol. 26, No. 3, pp. 317-30.

Nasar, Sylvia. 1998a. "New Breed of College All-Star: Columbia Pays Top Dollar for Economics Heavy Hitter." *New York Times* (8 April): p. D 1.

———. 1998b. *A Beautiful Mind: A Biography of John Forbes Nash, Jr.* (NY: Simon & Schuster).

National Science Board. 2000. *Science and Engineering Indicators, 2000* (Washington, D.C.: National Science Federation). <http://www.nsf.gov/sbe/srs/seind00/>

Nau, Henry. 1974. *National Politics and International Technologye Nuclear Reactor Development in Western Europe* (Baltimore: Johns Hopkins Press).

Nelson, Richard R. 1962. "The Link Between Science and Invention: The Case of the Transistor." in Richard R. Nelson, ed. *The Rate and Direction of Inventive Activity: Economic and Social Factors* (Princeton: Princeton University Press): pp. 549-83; reprinted in Richard R.

Nelson. 1996. *The Sources of Economic Growth* (Cambridge: Harvard University Press): pp. 159-88.

———. 1982. "The Role of Knowledge in R&D Efficiency." *Quarterly Journal of Economics,* Vol. 97, No. 3 (August): pp. 453-470; reprinted in Richard R. Nelson. 1996. *The Sources of Economic Growth* (Cambridge: Harvard University Press): pp. 147-58.

———. 1990. "Capitalism as an Engine of Progress." *Research Policy,* pp. 193-214; reprinted in Richard R. Nelson. 1996. *The Sources of Economic Growth* (Cambridge: Harvard University Press): pp. 52-82.

———. 1996. *The Sources of Economic Growth* (Cambridge: Harvard University Press).

Nelson, Richard and Nathan Rosenberg. 1994. "American Universities and Technical Advance in Industry." *Research Policy,* Vol. 23, No. 3 (May): pp. 323-348.

Nelson, Richard and Sidney Winter. 1982. *An Evolutionary Theory of Economic Change* (Cambridge, MA: Belknap Press).

Newcomb, Peter. 1999. "The Richest People in America." *Forbes* (11 October): p. 169.

Newman, Nathan. 1999. "Trade Secrets and Collective Bargaining: A Solution to Resolving Tensions in the Economics of Innovation." unpublished.

Nieves, Evelyn. 2000a. "Many in Silicon Valley Cannot Afford Housing, Even at $50,000 a Year." *New York Times* (20 February): p. A 20.

———. 2000b. "In San Francisco, Renters are Humble Supplicants." *New York Times* (6 June).

Noble, David. 1979. *America by Design* (Oxford University Press).

Nordhaus, William D. 1969. "An Economic Theory of Technological Change (in Theory of Innovation)." *American Economic Review,* Vol. 59, No. 2 (May): pp. 18-28.

O'Brien, Tia. 2001. "Top VC Doerr Apologizes for Helping Fuel Dot-Com Frenzy." *San Jose Mercury News* (16 July).

O'Connor, Michael J. L. 1944. *The Origins of Academic Economics in the United States* (NY: Columbia University Press).

Ohmann, Richard. 2000. "Historical Reflections on Accountability." *Academe* (January-February): pp. 24-29.

Orlik, Peter. 2001. "American Society of Composers, Authors and Publishers (ASCAP)." *Encyclopedia of Radio* (Chicago and London: Fitzroy Dearborn Publishers, forthcoming). <http://www.fitzroydearborn.com/chicago/radioascap.htm>

Paterson, Tony. 1999. "US Spy Satellites 'Raiding German Firms' Secrets'." *Electronic Telegraph* (11 April). http://www.telegraph.co.uk/et?ac=001545599564784&rtmo=kJAY3x3p&atmo=oooolsb&pg=/et/99/4/11/wspy11.html.

Pendergrast, Mark. 1999. *Uncommon Grounds: The History of Coffee and How It Transformed Our World* (NY: Basic Books).

Penrose, Edith T. 1951. *The Economics of the International Patent System* (Baltimore: Johns Hopkins University Press).

———. 1959. *The Theory of the Growth of the Firm* (Oxford: Basil Blackwell).

Pepper, Robert [Chief, FCC Office of Plans and Policy]. 1995. Letter to Senator Joseph Lieberman (6 September).

Perelman, Michael A. 1991. *Information, Social Relations, and the Economics of High Technology* (NY and London: St. Martin's and Macmillan, 1991).

———. 1996. *The End of Economics* (London: Routledge).

———. 1998. *Class Warfare in the Information Age* (NY: St. Martin's Press).

———. 1999. *The Natural Instability of Markets: Expectations, Increasing Returns and the Collapse of Markets* (NY: St. Martin's Press).

———. 2000. *Transcending the Economy: On the Potential of Passionate Labor and the Wastes of the Market* (NY: St. Martin's Press).

Peters, Tom. 1997. *The Circle of Innovation* (NY: Alfred A. Knopf).

Petersen, Melody. 2000a. "What's Black and White and Sells Medicine?" *New York Times* (27 August).

Petersen, Melody. 2000b. "Pushing Pills With Piles Of Money." *New York Times* (5 October).

Philipkoski, Kristen. 2000. "Genetics Scandal Inflames Iceland." *Wired News* (March 20). <http:/ /www.wired.com/news/politics/0,1283,35024,00.html>

Philips, Chuck. 2000. "Time Warner Tunes in New Delivery Channel." *Los Angeles Times* (25 July).

Plant, Arnold. 1934. "The Economic Theory Concerning Patents for Inventions." *Economica,* ns. Vol. 1, No. 1 (February): pp. 30-51.

Plotkin, Mark J. 1993. *Tales of a Shaman's Apprentice: An Ethnobotanist Searches for New Medicines* (NY: Penguin).

Polanvyi, Michael. 1944. "Patent Reform." *Review of Economic Studies,* Vol. 11, No. 2 (Summer): pp. 61-76.

Pollack, Andrew. 1999. "Biological Products Raise Genetic Ownership Issues." *New York Times* (26 November).

———. 2000a. "Two Paths to the Same Protein." *New York Times* (28 March).

———. 2000b. "Columbia Gets Help From Alumnus on Patent Extension." *New York Times* (21 May): p. A 37.

———. 2000c. "U.S. Hopes to Stem Rush Toward Patenting of Genes." *New York Times* (28 June).

———. 2000d. "Amgen Wins Court Battle Over Drug For Anemia." *New York Times* (20 January).

———. 2001. "The Green Revolution Yields to the Bottom Line." *New York Times* (15 May).

Port, Otis. 1984. "Can Bell Labs' Magic Survive the AT&T Breakup?" *Business Week* (8 October): p. 19.

Post, David G. [Temple University Law School/Cyberspace Law Institute] 1998. *Some Thoughts On The Political Economy Of Intellectual Property: A Brief Look at the International Copyright Relations of the United States* <http://www.nbr.org/regional_studies/ipr/chongqing98/ post_essay.html>

Powell, Irene. 1987. "The Effect of Reductions in Concentration on Income Distribution." *Review of Economics and Statistics,* Vol. 49, No. 1 (February): pp. 75-82.

Powell, W. W. and J. Owen-Smith. 1998. "Universities and the Market for Intellectual Property in the Life Sciences." in Burton A. Weisbrod, ed. *To Profit or Not to Profit: The Commercial Transformation of the Nonprofit Sector* (Cambridge: Cambridge University Press, 1998): pp. 169-93.

Press, Eyal and Jennifer Washburn. 2000. "The Kept University." *Atlantic Monthly,* Vol. 285, No. 3 (March): pp. 39-54.

Pugh, Emerson W. 1984. *Memories That Shaped an Industry* (Cambridge: MIT Press).

———. 1995. *Building IBM: Shaping an Industry and its Technology* (Cambridge: MIT Press).

Rajan, Raghuram G. and Luigi Zingales. 2001. "The Firm as a Dedicated Hierarchy: A Theory of the Origins and Growth of Firms." *Quarterly Journal of Economics,* Vol. 116, No. 3 (August): pp. 805-52.

Rand, Ayn. 1964. "Patents and Copyrights." *Objectivist Newsletter* (May); reprinted in Ayn Rand. *Capitalism: The Unknown Ideal* (NY: New American Library, 1966): pp. 125-29.

Reed, Deborah. 1999. *California's Rising Income Inequality: Causes and Concerns* (San Francisco: Public Policy Institute of California). <http://www.ppic.org/publications/PPIC116/ index.html>.

Regalado, Antonio. 2000. "The Great Gene Grab." *Technology Review* (September/October): pp. 48-55.

———. 2001. "Tiny TechSearch Wields Patents Against Giant-Sized Competitors." *Wall Street Journal* (9 March).

Reich, Leonard S. 1977. "Research, Patents, and the Struggle to Control Radio." *Business History Review,* Vol. 11, No. 2 (Summer): pp. 208-35.

Reppy, Judith. 1985. "Military R&D and the Civilian Economy." *Bulletin of the Atomic Scientists,* Vol. 41, No. 9 (October): pp. 10-14.

Richards, Bill. 1999. "Baxter Beat CellPro in Court, But Some Say Dying Patients Lose." *Wall Street Journal* (6 August): p. A 1.

Ridge, Pamela Sebastian. 2000. "Business Bulletin: Aging Cats and Dogs Get More Attention From Drug." *Wall Street Journal* (16 November): p. A 1.

Rivette, Kevin G. and David Kline. 2000. *Rembrandts in the Attic: Unlocking the Hidden Value of Patents* (Boston: Harvard Business School Press).

Robbins, Lionel Charles. 1939. *The Economic Basis of Class Conflict and Other Essays* (London: Macmillan).

———. 1969. *An Essay on the Nature and Significance of Economic Science,* 2d ed. (London: Macmillan).

Roberts, Leslie. 1991. "Genome Patent Fight Erupts." *Science,* Vol. 254 (11 October): pp. 184-86.

Rosenbaum, David E. 1999. "The Gathering Storm Over Prescription Drugs." *New York Times* (14 November): p. D 1.

Rosenthal, Elisabeth. 1993. "Research, Promotion and Profits: Spotlight Is on the Drug Industry." *New York Times* (February 21): p. A1.

Ross, Emma. 2000. "British Scientists Object To Utah Company's Gene Patent Attempt." *Associated Press* (19 January).

Ryan, Michael P. 1998. *Knowledge Diplomacy: Global Competition and the Politics of Intellectual Property* (Washington, D.C.: Brookings Institution Press).

Sampat, Payal. 1999. "Editorial: The New Flat Earth Society." *World-Watch Magazine* (November/December).

Samuelson, Pamela. 1999. "Intellectual Property and the Digital Economy: Why the Anti-Circumvention Regulations Need to Be Revised." *Berkeley Technology Law Review,* Vol. 14, No. 2 (Spring).

Samuelson, Paul A. 1954. "The Pure Theory of Public Expenditures." *Review of Economics and Statistics,* Vol. 36, pp. 387-99.

———. 1958. "Aspects of Public Expenditure Theories." *Review of Economics and Statistics,* Vol. 40, (November): pp. 332-6.

———. 1964. "Public Goods and Subscription TV." *Journal of Law and Economics,* Vol. 7 (October): pp. 81-83.

———. 1969. "Pure Theory of Public Expenditure and Taxation." in J. Margolis and H. Guitton, eds., *Public Economics: An Analysis of Public Production and consumption and Their Relations to the Private Sector* (London: St. Martin's Press): pp. 98-123.

Sandburg, Brenda. 1999a. "Patent Applications Flow Freely." *Legal Times* (22 February): p. 12.

———. 1999b. "Speed Over Substance." *Intellectual Property Magazine* (March).

Savio, Mario. 1964. "Speech," (3 December). <http://www.fsm-a.org/stacks/mario/mario_speech.html>

Saxenian, AnnaLee. 1994. *Regional Advantage: Culture and Competition in Silicon Valley and Route 128* (Cambridge: Harvard University Press).

———. 1999. "Comment on Kenney and von Burg, 'Technology, Entrepreneurship and Path Dependence: Industrial Cluster in Silicon Valley and Route 128'." *Industrial and Corporate Change,* Vol. 8, No. 1 (March): pp. 105-10.

Scherer, Frederick M. 1950. *Capitalism, Socialism and Democracy,* 3d. ed. (New York: Harper & Row).

———. 1980. *Industrial Market Structure and Economic Performance,* 2nd. ed. (Chicago: Rand McNally).

———. 1996. *Industry Structure, Strategy and Public Policy* (NY: Harper Collins).

Schiller, Herbert I. 1985. "Supply-Side Knowledge: Information—A Shrinking Resource." *The Nation* (28 December): p. 708.

Schoofs, Mark. 2000. "Ebola Researchers Report Progress, But a Long Road Remains Ahead." *Wall Street Journal* (30 November): p. B 1.

Schumpeter, Joseph Alois. 1950. *Capitalism, Socialism and Democracy,* 3d. ed. (New York: Harper & Row).

Sell, Susan K. 1999. "Multinational Corporations as Agents of Change: The Globalization of Intellectual Property Rights." in *Private Authority and International Affairs,* eds. A. Claire Cutler, Virginia Haufler, and Tony Porter (Albany: State University of New York Press): pp. 169-98.

Shaffer, Marjorie. 1992. "Survey of Pharmaceuticals." *Financial Times* (23 July): p. 29.
Shane, Scott and Tom Bowman. 1995. "America's Fortress of Spies." *Baltimore Sun* (3 December).
Shea, Christopher. 2000. "No Tenure, No Peace." *Lingua Franca*, Vol. 10, No. 8 (November).
Shear, Michael D. and Tom Jackman. 2000. "Fairfax Boom Leaves Little Room for Poor." *Washington Post* (14 March 14): p. A1.
Shiller, Robert J. 2000. *Irrational Exuberance* (Princeton: Princeton University Press).
Shulman, Seth. 1999. *Owning the Future: Staking Claims on the Knowledge Frontier* (Boston: Houghton Mifflin).
———. 2000. "Software Patents Tangle the Web." *Technology Review* (March/April).
———. 2001a. "IP's Bleak House." *Technology Review* (March/April).
———. 2001b. "PB&J Patent Punch-up." *Technology Review* (May).
———. 2001c. "Looting the Library." *Technology Review* (June): p. 37.
Silverstein, Ken. 1999. "Millions for Viagra, Pennies for Diseases of the Poor." *The Nation* (19 July).
Siwek, Stephen E. and Gale Mosteller. 1996. Copyright Industries in the U.S. Economy: The 1996 Report (Economists Incorporated). <http://www.iipa.com/html/pn_executive_summary.html>
Slaughter, Sheila and Larry Leslie. 1997. *Academic Capitalism: Politics, Policies and the Entrepreneurial University* (Baltimore: John Hopkins).
Slawson, David. 1981. *The New Inflation: The Collapse of Free Markets* (Princeton: Princeton University Press).
Slind-Flor, Victoria. 1999. "Stalking the Submarine Patent King: Will the Enemies of Jerome Lemelson Have the Last Laugh?" *IP Worldwide* (October).
Smith, Adam. 1776. *The Nature and Causes of the Wealth of Nations* (Oxford: Oxford University Press, 1976).
Smith, G. Kenneth. 1997. "Faculty and Graduate Student Generated Inventions: Is University Ownership a Legal Certainty?" *Virginia Journal of Law and Technology*, Vol. 1, No. 4 (Spring). <http://vjolt.student.virginia.edu/graphics/vol1/home_art4.html>
Snoddy, Raymond. 1999. "Corporate Profile [of Reed Elsevier]." *Times of London* (4 January): p. 44.
Soley, Lawrence C. 1995. *Leasing the Ivory Tower: The Corporate Takeover of Academia* (Boston: South End Press).
Specter, Michael. 1999. "Decoding Iceland." *New Yorker* (18 January): pp. 40-53.
Spink, Lynn. 1997. "Right to Poverty: Fear and Loathing at a Fraser Institute Right to Work Conference." *This Magazine* (Canada) 30: 4 (January/February): pp. 20-21.
Standard Oil Co. (Ind.) v. United States. 1931. 283 U.S. 163, 167-68.
Stanecki, Karen A. 2000. "The AIDS Pandemic in the 21st Century: The Demographic Impact in Developing Countries." Paper at the 13th International AIDS Conference, Durban, South Africa (9-14 July).
Steckel, Francis C. 1990. "Cartellization of the German Chemical Industry, 1918-1925." *Journal of European Economic History*, Vol. 19, No. 2 (Fall): pp. 329-51.
Stelfox, Henry Thomas, Grace Chua, Keith O'Rourke, and Allan S. Detsky. 1998. "Conflict of Interest in the Debate over Calcium-Channel Antagonists." *The New England Journal of Medicine*, Vol. 338, No. 2 (8 January): pp. 101-6.
Stephan, Paula E. 1996. "The Economics of Science." *Journal of Economic Literature*, Vol. 34, No. 3 (September): pp. 1199-1262.
Stephan, Paula E. and Sharon G. Levin. 1992. *Striking the Mother Lode in Science: The Importance of Age, Place, and Time* (New York: Oxford University Press),
Stephens, Joe. 2000. "The Body Hunters: As Drug Testing Spreads, Profits and Lives Hang in Balance." *Washington Post* (17 December): p. A01.
Stern, Bernard J. 1949. "The Corporations as Beneficiaries." *American Scholar*, Vol. 28.
Stern, Scott. 1999. "Do Scientists Pay to Be Scientists?" National Bureau of Economic Research Working Paper No. 7410 (October).
Stix, Gary. 2001a. "The Mice That Warred." *Scientific American*, 284: 6 (June): pp. 35-35.
Stix, Gary. 2001b. "Code of the Code." *Scientific American*, 284: 6 (June): p. 32.

Stolberg, Sheryl Gay and Jeff Gerth. 2000. "Holding Down the Competition: How Companies Stall Generics and Keep Themselves Healthy." *New York Times* (23 July).

Strohm, John W. 1999. "Between Academia and Industry." *The Tech,* Vol. 119, No. 29 (7 July). < http://www-tech.mit.edu/V119/N29/col29stroh.29c.html>.

Sturgeon, Timothy J. 2000. "How Silicon Valley Came to Be." in Martin Kenney, ed. *Understanding Silicon Valley: The Anatomy of an Entrepreneurial Region* (Stanford: Stanford University Press): pp. 15-47.

Sturmey, S. G. 1958. *The Economic Development of Radio* (London: Duckworth).

Takahashi, Dean. 1999. "Intel's Efforts to Thwart a Rival Criticized as Unethical." *Wall Street Journal* (19 April).

Teitelman, Robert. 1994. *Profits of Science: The America Marriage of Business and Technology* (NY: Basic Books).

Ten Kate, Kerry and Sarah A. Laird. 1999. *The Commercial Use of Biodiversity: Access to Genetic Resources and Benefit-Sharing* (London: Earthscan).

Tocqueville, Alexis de. 1835. *Democracy in America.* Philip Bradley, ed. (New York: Vintage, 1945).

Uhrhammer, Jerry. 1999. "Wasabi's Hot Stuff For Oregon Growers." *Portland Oregonian* (March 16).

United Nations Development Programme. 1999. *Globalization with a Human Face: United Nations Human Development Report* (NY: Oxford University Press).

United States Congress, Office of Technology Assessment. 1992. *Federal and Private Roles in the Development and Provision of Alglucerase Therapy for Gaucher Disease,* OTA-BP-H-104 (Washington, DC: U.S. Government Printing Office, October).

———. 1993. *Pharmaceutical R&D: Costs, Risks and Rewards,* OTA-H-522 (Washington, DC: U.S. Government Printing Office, February).

United States Department of Agriculture. National Agricultural Statistics Service. 2000. *Agricultural Statistics 2000* (Washington, D.C.: U.S. Government Printing Office). <http://www.usda.gov/nass/pubs/agr00>

United States Department of Housing and Urban Development. 2000. *The State of the Cities, 2000.* <http://www.huduser.org/publications/pdf/socrpt.pdf>

United States Patent and Trademark Office. 1999. *Annual Report, 1998.* <http://www.uspto.gov/web/offices/com/annual/1998/a98r-1.htm#Topic9>.

Usselman, Steven W. 1999. "Patents, Engineering Professionals, and the Pipelines of Innovation: The Internalization of Technical Discovery by Nineteenth Century American Railroads." in *Learning by Doing in Markets, Firms, and Countries,* Naomi R. Lamoreaux, Daniel M. G. Raff, and Peter Temin, eds. (Chicago: University of Chicago Press): pp. 64-91.

Valenti, Jack (Chairman and Chief Executive Officer, Motion Picture Association). 1999. "Statement Before the Committee on Ways and Means Subcommittee on Trade, Regarding US-China Trade Relations and the Possible Accession of China to the World Trade Organization (8 June)." <www.mpaa.org/jack/99/99_6_8a.htm>

Vanderbilt, Tom. 1998. *The Sneaker Book: Anatomy of an Industry and an Icon* (NY: New Press).

Varian, Hal. 1993. "Markets for Public Goods." *Critical Review,* Vol. 7, No. 4 (Fall): pp. 539-57.

Vaughan, Floyd Lamar. 1972. *The United States Patent System: Legal and Economic Conflicts in American Patent History* (Westport, CT: Greenwood Press).

Veblen, Thorstein. 1915. *Imperial Germany and the Industrial Revolution* (New York: Macmillan & Co.).

———. 1918. *The Higher Learning in America: Memorandum on the Conduct of Universities by Business Men* (NY: B. W. Huebsch).

Vickers, Marcia. 2001. "When Wealth Is Blown Away." *Business Week* (26 March): pp. 39-41.

Vogel, Gretchen. 1997. "Long-Suppressed Study Finally Sees Light of Day." *Science,* Vol. 676, No. 2312 (25 April 1997): pp. 525-26.

von Hippel, Eric. 1988. *The Sources of Innovation* (Cambridge: MIT Press).

Voss, David. 1999. "'New Physics' Finds a Haven at the Patent Office." *Science* (21 May): pp. 1252-54.

Waldman, Peter. 1999. "In Silicon Valley, The Conversation Comes with a Nondisclosure Form." *Wall Street Journal* (3 November): p. A 1.

Walsh, Mark. 1996. "Patently Ridiculous?" *Intellectual Property Magazine* (October).

———. 1997. "Patently Ridiculous." *Internet World Magazine* (April).

Warshofsky, Fred. 1994. *The Patent Wars: The Battle to Own the World's Technology* (NY: Wiley).

Washburn, Jennifer. 2001. "Undue Influence." *American Prospect,* Vol. 12, No. 14 (August). <http://www.prospect.org/print-friendly/print/V12/14/washburn-j.html>

Webber, Tom. 1999. *Wall Street Journal* "Battles Over Patents Could Hurt the Web." (8 November): p. B 1.

Weiner, Rebecca S. 2000. "Computer Science Departments Are Depleted as More Professors Test Entrepreneurial Waters." *New York Times* (9 August).

Wernick, Iddo K., Robert Herman, Shekhar Govind, and Jesse H. Ausubel. 1996. "Materialization and Dematerialization: Measures and Trends." *Daedalus,* Vol. 125, No. 3 (Summer): pp. 171-198. <http://phe.rockefeller.edu/Daedalus/Demat/>

White, Ed. 2000. "No Decision On GMO Case Until Fall." *Western Producer* (29 June). <http://www.producer.com/articles/20000629/news/20000629news04.html>

Wilke, John R. and James Bandler. 2001. "New Digital Camera Deals Kodak a Lesson in Microsoft's Methods: Trial Use With Windows XP Gave Microsoft an Edge, Photo Firm Says." *Wall Street Journal* (2 July): p. A 1.

Williams, Daniel. 2000. "New Russian Firm Reinvents the Bottle." *Washington Post* (30 July): p. A 6.

Williams, James C. 1998. "Frederick E. Terman and the Rise of Silicon Valley." *International Journal of Technology Management,* Vol. 16, No. 8, pp. 751-60.

Willman, David. 2000. "How a New Policy Led to Seven Deadly Drugs Medicine: Once a Wary Watchdog, the U.S. Food and Drug Administration Set Out to Become a 'Partner' of the Pharmaceutical Industry." *Los Angeles Times* (20 December).

Winter, Sidney G. 1982. "An Essay on the Theory of Production." in S. H. Hymans, ed. Economics and the World Around It (Ann Arbor: University of Michigan Press): pp. 55-91.

———. 1991. "On Coase, Competence, and the Corporation." *The Nature of the Firm* in Oliver E. Williamson and Sidney G. Winter, eds. (Oxford: Oxford University Press): pp. 179-95.

Wolpert, Lewis and Alison Richards. 1988. *A Passion for Science* (New York: Oxford University Press).

Wong, May. 2000. "High-Stakes Battle Waged Over Patents for Internet Techniques, Business Methods Law: Companies Hope for Lucrative Payoff by Laying Legal Claim to Such Commonplace Features as Clicking to Jump from One Web Site to Another." *Sacramento Bee* (19 July): p. C 1.

Woolsey, R. James. 2000. "Why We Spy on Our Allies." *Wall Street Journal* (17 March).

World Health Organization. 2000. *World Health Report 2000* (Geneva: World Health Organization).

Wysocki, Bernard Jr. 2000. "In U.S. Trade Arsenal, Brains Outgun Brawn." *Wall Street Journal* (10 April): p. A 1.

Yerton, Stewart. 1993. "The Sky's The Limit." *The American Lawyer* (May): p. 64.

Zachary, G. Pascal. 1997. *Endless Frontier: Vannevar Bush, Engineer of the American Century* (NY: Free Press).

Zacks, Rebecca. 2000. "The TR University Research Scorecard." *Technology Review* (July/August).

Zamel, Noe. 1995. "In Search of the Genes of Asthma on the Island of Tristan da Cunha." *Canadian Respiratory Journal,* Vol. 2, No. 1, pp. 18-22.

Zitrin, Richard and Carol M. Langford. 2000. "The Moral Compass: Law Firms Seem to be Stealing Signs from Baseball on Salary Issues." *American Lawyer Media* (4 February). <http://www.lawnewsnetwork.com/stories/A15267-2000Feb4.html>.

Zolla-Pazner, Susan. 1994. "The Professor, the University, and Industry." *Scientific American,* Vol. 268, No. 3 (March): p. 120.

Zorpette, Glenn. 2000. "All Doped Up and Going for the Gold." *Scientific American,* Vol. 282, No. 5 (May): pp. 20-22.

INDEX

A

B

D

H

I

J

K

L

N

O

P

Q

R

S

T

U

V